Powering the Future: Tailoring Organic Semiconductors for Efficient Devices

Herlie

Contents

Contents

1. Introduction

The field of organic molecular semiconductors constitutes a conjunction of molecular and solid states physics incorporating compounds with few electronically bound atoms into aggregates consisting of a large amount of building blocks and, interestingly, showing new functionalities but still preserving many properties of the individual molecular entities. Molecules bound in a crystal are subject to interactions with their neighboring partners while keeping their local dynamics, *e.g.* vibrational excitations, leading to emergent phenomena such as virtually forbidden electronic transitions at the edge of the Brillouin zone enabled by generation of a molecular vibration, hence, conserving the momentum. A particularity of molecular crystals originates from the size of the molecules constituting its building blocks. The interaction between the crystal's constituents is not only determined by their arrangement but by their respective orientation as well. Probably, the best known photophysical phenomenon resulting from the crystal structure and the associated molecular packing is the striking blue or red shift of the absorption wavelengths due to the so-called *H-* or *J-type* molecular packing yielding also an enhanced photon emission of the latter and a suppressed luminescence of the former [Spa10][GP13][SB16]. This luminescence enhancement can even reach *superradiant* characteristics in highly ordered aggregates [SY11][Mül+13][Eis+17]. The implications of the molecular orientation on the physical properties are not limited to purely academic interest, but are of relevance for applications in organic electronic devices, too. For example, the formation of charge transfer states as well as the subsequent charge separation in *donor-acceptor* (D-A) heterojunctions is highly dependent on the molecular orientation at the D-A interface, thus, crucially determining the efficiency of organic photovoltaic cells [Ran+12][Agh+14][Ran+17]. Therefore, this work aims to elucidate the influence of molecular orientation and aggregate packing on photo excited states of technologically relevant molecular materials as well as at donor-acceptor interfaces. Moreover, possible strategies to exploit the control over the molecular packing to specifically tune the photophysical behavior of prototypical devices are analyzed.

With the idea in mind to investigate the formation of charge transfer states as function of the donor-acceptor orientation the prototypical charge transfer (CT) complex pentacene-perfluoropentacene (PEN-PFP) was employed, known for its strong CT interaction in mixed crystalline films [Bro+11][Ang+12]. The planaritiy of the molecules suggests a strong spatial anisotropy of the CT formation. To overcome structural imperfections inherent to crystalline films, PEN single crystals were utilized as substrates of highest structural quality for PFP film growth enabling the investigation of CT formation right at the donor-acceptor single crystal interface by means of highly sensitive optical reflection and photoluminescence spectroscopy.

A conventional approach to investigate the impact of the molecular packing on the aggregate photophysics is to chemically modify the side groups of a common molecular backbone, thus, modifying the van der Waals radii of the molecular constituents and hence, tuning the molecular packing [GP13]. Even though highly successful, this technique bears a large expense in synthesis as well as the danger of disturbing the electronic system of the molecular backbone and thereby, impeding the comparability between different molecular aggregates. To overcome these shortcomings, zinc phthalocyanine (ZnPc) was used by virtue of its polymorphism making the fabrication of J- and H-aggregates feasible by controlling the crystallization conditions [Erk04][Luc+20] while keeping the molecular orbitals conserved. Using temperature dependent photoluminescence the photophysical properties and their relation to the molecular packing are examined over a large temperature range between 5 K and 300 K. The emission is analyzed using either an excimer description based on a quantum mechanical harmonic oscillator model or an exciton polaron description [Spa10] suitable to describe both, the Frenkel exciton emission at room temperature as well as the superradiant emission at cryogenic temperatures of the analyzed ZnPc single crystals. Moreover, by measuring photoluminescence during thermal annealing of ZnPc thin films, the kinetics of the phase transition between the two polymorphs can be traced by virtue of the spectrally distinct emission processes which, *vice versa*, can be used to tune the spectral emission of such layers.

In the first part of this work, chapter 2 gives a summary of the theoretical basics of crystal growth, electronic transitions and excited states in molecules and molecular aggregates, and on organic light emitting diodes needed in this work. Moreover, chapter 3 introduces the examined materials and various experimental techniques for structural and optical characterization. This theoretical introduction is followed by two chapters presenting the main experimental results. Chapter 4 reports on the spatially anisotropic CT state formation in Pentacene-Perfluoropentacene and its impact on the device performance of two types of prototypical heterojunction diodes. Chapter 5 completes this book by a complementary description of the molecular packing and the resulting photophysics of two zinc phthalocyanine polymorphs.

2. Theoretical Background

2.1. Organic Semiconductors

The material class of organic semiconductors is based on hydrocarbons and can be roughly divided into molecules and polymers. While polymers are easily processed from solution they mostly lack long range ordering, wheras small molecules can form highly crystalline aggregates making them less sensitive to disorder induced disturbances of their physical properties. Although the class of polymers is widely studied for application their inherent disorder shifts them out of the focus of this work, which is why the following theoretical concepts and aspects mostly concentrate on small molecules. The optical and electronic properties are determined by the characteristics of the typical hydrocarbon π-electron system, while the mechanical properties of the formed solid states result from van der Waals binding between the molecules.

2.1.1. Hybrid Orbitals and Molecular Bonds

At the core of organic semiconductor's physics lies the hybridization of the carbon atom orbitals. The carbon atom, in its ground state, has six electrons, fully occupying the 1s and 2s orbitals as well as half occupying two of the three degenerated 2p orbitals as shown in figure 2.1 a). To explain directional bonding, the three degenerated canonical 2p orbitals with magnetic quantum number $m_l = -1, 0, 1$ are usually recombined to wave functions which have an orthogonal representation in space [MS97]. Figure 2.1 b) shows surfaces of equal probability as a representation of these atomic orbitals. For simplification, instead of talking about "one representation of the atomic (or molecular) orbital as a surface of equal probability", the expression "atomic/molecular orbital" will be used when describing a graphical representation of such orbitals. It is evident that each formed p orbital is aligned along one distinguished spatial direction, in contrast to the isotropic s orbital shown in Figure 2.1 c). The orbitals are referred to as p_x, p_y, and p_z, respectively. The corresponding wave function has positive real values for $x, y, z > 0$ and negative values for $x, y, z < 0$, exemplary visualized for p_z in Figure 2.1 b).

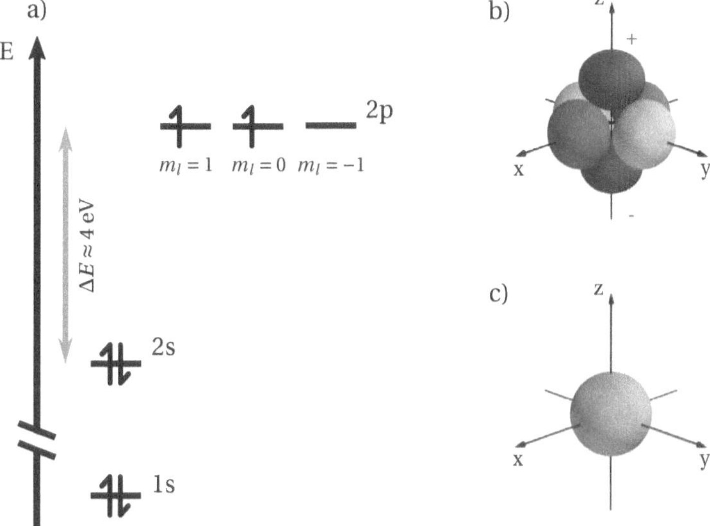

Figure 2.1.: a) Ground state configuration and schematic energy levels of the carbon atom. Promotion energy for a 2s electron is approximately 4 eV [Moo70]. b) Representation of 2p orbitals as surfaces of equal probability. Direction dependency is achieved through proper linear combination of canonical $n = 2$, $l = 1$ wave functions. c) 2s orbital presented as equal probability surface.

The energetic difference between the 2p orbitals and the 2s orbital is 4.19 eV and an excited state with one electron from the 2s state promoted to the empty 2p state exists [Moo70]. In the presence of another atom nearby the necessary promotion energy can be compensated by the energetically favorable molecular binding [Dem10] and in the excited state the 2s orbital and the 2p orbitals mix, causing the formation of new states based on linear combinations of the initial ones [HW03]. This process is generally know as *hybridization* of atomic orbitals and the result is called sp^1, sp^2, or sp^3 *hybridization* depending on the number of participating p orbitals. Figures 2.2 a), c) and e) show the resulting orbitals as well as their schematic energy levels and electron occupation. The resulting wave functions are orthogonal in a quantum mechanical sense, but, as is evident, show a distinct geometrical configuration which is characteristic for each hybridization.

The sp^1 hybridization leads to two orbitals along the z direction. Each orbital shifts the electron density either in z or in -z direction as a result of the either constructive superposition of the s and p_z orbital along the z direction for $|sp_{+z}\rangle = 1/\sqrt{2}(|s\rangle + |p_z\rangle)$ or the destructive superposition in case of $|sp_{-z}\rangle = 1/\sqrt{2}(|s\rangle - |p_z\rangle)$ [MS97]. The two resulting hybrid orbitals are shown in Figure 2.2 a) (purple) together with the two remaining p orbitals (grey) as well as their schematic energy levels and electron occupation. This type of hybridization is responsible for the carbon triple bond found, *e.g.* in ethyne, where the hybrid orbitals form the C–C and C–H σ bondings, and the remaining carbon p orbitals form two degenerated π orbitals, oriented perpendicular to each other and stretched out along the molecular axes as shown in Figure 2.2 b) [EG08].

The sp^3 configuration, where the s and all p orbitals form four hybrid states, exhibits a tetrahedral structure in which every orbital can form a stable σ bond with neighboring atoms (see Fig. 2.2 e)). The methane molecule is formed with four hydrogen atoms, each bound to one hybrid orbital and, thus, sitting in the corners of the respective tetrahedral as shown in figure 2.2 f). [HW03] If each sp^3 oribtal forms a σ bond with another sp^3 hybridized C atom, a tightly bound carbon crystal is formed, famously known as diamond [Dem10].

For organic semiconductors, the most important configuration is the sp^2 hybridization [MS97]. The s, the p_x, and the p_y state form three hybrid orbitals which span out in the x-y plane, mutually enclosing 120° angles, while the remaining p_z orbital sticks out of the plane along the z direction. The orbitals are depicted together with their respective energy levels in Figure 2.2 c). Planar molecules are usually based on the sp^2 hybridized carbon atoms, where the hybrid orbitals form σ bonds with neighboring carbon or hydrogen atoms, while the p_z orbitals overlap and form a π bond along the carbon-carbon axis. If the number of carbon atoms is greater than two, the π-elec-

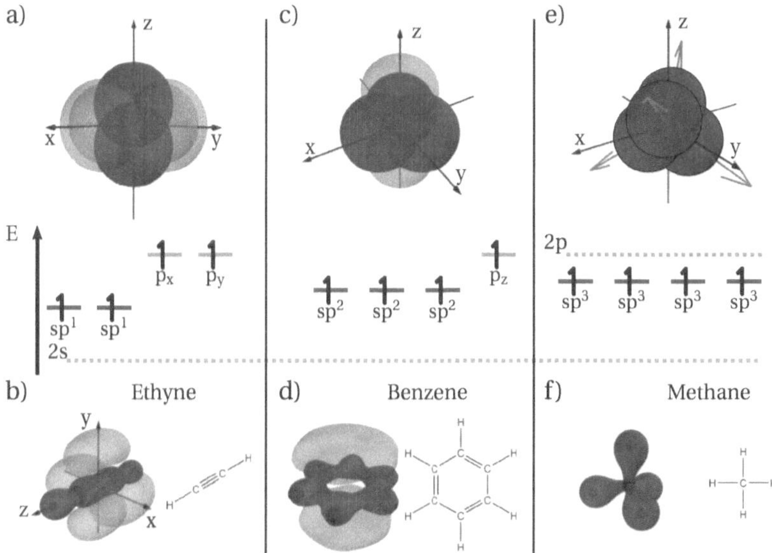

Figure 2.2.: Hybrid orbitals (purple) and remaining p orbitals (grey) after C atom hybridization with their respective energy levels and electron occupation as well as molecular examples for each type. The dotted green line marks the energy level of the 2s orbital for reference. a) Sp^1 hybrid orbitals aligned along the z axis together with the p_x and p_y orbitals lying in the x-y plane. b) The displayed ethyne molecule is an example for sp^1 hybridized C atoms forming a linear molecule along the z axis. c) The triangular configuration of the three sp^2 hybrid orbitals in the x-y plane and the remaining p_z orbital in z direction. d) the benzene ring is presented as an example composed of six sp^2 hybridized C atoms (H atoms not shown for clarity) and the delocalized conjugated π-electron system composed of the p_z orbitals. e) Tetrahedral configuration of the sp^3 hybridization highlighted by arrows. f) The methane molecule as a classical example for the sp^3 hybridization is shown.

tron system usually delocalizes, forming a *conjugated π-electron system* which will be discussed in its detailed properties in section 2.1.2. One key unit of many unsaturated hydrocarbons is benzene with its structure shown in figure 2.2 d). Six sp^2 carbon atoms make up a ring, where each carbon atom forms a σ bond to its two neighboring carbon atoms and to one hydrogen atom in the periphery (not shown in 3D illustration). The remaining p orbitals form a delocalized π-electron system below and above the plane defined by the carbon ring.

2.1.2. Conjugated π-systems

A conjugated π-electron system, commonly abbreviated *π-system*, results from sp^2 hybridized carbon atoms connected by alternating σ and overlapping σ-π bonds, leading to alternating single and double bonds. A simple example is given by the infinite polyene chain *trans*-polyacetylene $(C_2H_2)_n$, shown in figure 2.3.

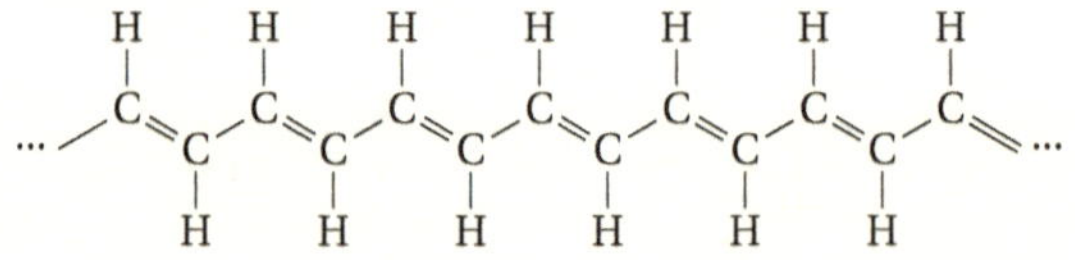

Figure 2.3.: The *trans*-polyacetylene polymer is a model system of the infinite polyene chain.

The p_z electron of a randomly picked carbon atom in the chain can either form an overlapping π orbital with the p_z electron of the carbon atom to the right or with the carbon atom to the left. Taking into account only the interaction with the next nearest neighboring C-atoms, in a simple approximation, the electron is moving in the potential of periodically arranged C-atoms in one dimension. This picture is well known in solid state physics, describing nearly free electrons in crystalline metals or semiconductors. There, the electron is treated as a delocalized particle and the Bloch ansatz leads to the formation of electronic bands [Dem10]. This approach can be easily extended to describe the π-electrons of *trans*-polyacetylene [BMA16]. This delocalization of p_z-electrons along the sp^2 hybridized carbon chain and the resultant formation of electronic bands is the core feature of the conjugated π-system, leading to the stabilization of molecules and determines their optical and electronic features. The 2000 Nobel Prize in Chemistry was awarded "for the discovery and development of conductive polymers" [Nob00].

For small molecules, *i.e.* carbon chains with a finite length or closed cycles, such as benzene (see Fig 2.2 d)), usually a method based on the linear combination of atomic

<u>o</u>rbitals (LCAO for short) of the p_z orbitals is used, which leads to a qualitative under-standing of the π-system formation. The *Hückel-method* or *Hückel molecular orbital theory*, introduced by Erich Hückel in 1930 [Hüc30], treats the σ orbitals as a scaffold determining the fixed positions of the respective p orbitals [MS97]. For the qualitative description of the molecules, the following simplifications of the molecular Hamilto-nian are assumed [FK83]:

- Only the energy of the π electrons is taken into account. Atomic nuclei and σ electrons are ignored.

- Only nearest neighbor interactions are considered.

- The Coulomb integrals over all atomic orbitals φ_i are set equal to the free param-eter $\alpha = \int \varphi_i \hat{H}_C \varphi_i dV < 0$, with $\hat{H}_C$ being the Coulomb interaction operator and are not calculated explicitly.

- The resonance integral of two neighboring orbitals φ_i and φ_j is set equal to the free parameter $\beta = \int \varphi_i \hat{H}_C \varphi_j dV < 0$ and is not calculated explicitly as well.

- The overlap integral $S_{ij} = \int \varphi_i \varphi_j dV$ is neglected.

For the p orbitals of the ethene molecule's two C atoms shown in figure 2.4 a), the Hückel-method yields a bonding π orbital and an antibonding π^* orbital given by the linear combinations

$$|\pi\rangle = \frac{1}{\sqrt{2}} \left(|p_1\rangle + |p_2\rangle \right) \tag{2.1}$$

and

$$|\pi^*\rangle = \frac{1}{\sqrt{2}} \left(|p_1\rangle - |p_2\rangle \right). \tag{2.2}$$

The respective energies are $E_\pi = \alpha - \beta$ and $E_{\pi^*} = \alpha + \beta$. The corresponding energy diagram and a schematic visualization of the linear combinations are shown in Figure 2.4 b). The π-π^* splitting determines most of the optical and electrical properties of small organic molecules. Electronic transitions usually occur between the fully occu-pied π orbital and the unoccupied π^* orbital. The π-π^* energy gap lies, in the majority of cases, in the range between 1 eV to 4 eV. On the contrary, the σ-σ^* gap of the under-lying σ scaffold usually amounts to around 10 eV [WH69].

For molecules consisting of more than two carbon atoms, an equal amount of bond-ing and antibonding orbitals is formed by the p_z orbitals. Benzene (see Fig. 2.2 d)), for example, has three fully occupied π orbitals of bonding character and three empty anti-bonding π^* orbitals. The occupied π orbital with the highest energy is refered to as the <u>h</u>ighest <u>o</u>ccupied <u>m</u>olecular <u>o</u>rbital, *HOMO* for short. The unoccupied π^* orbital with the lowest energy, *vice versa*, is termed *LUMO* - <u>l</u>owest <u>u</u>noccupied <u>m</u>olecular <u>o</u>rbital.

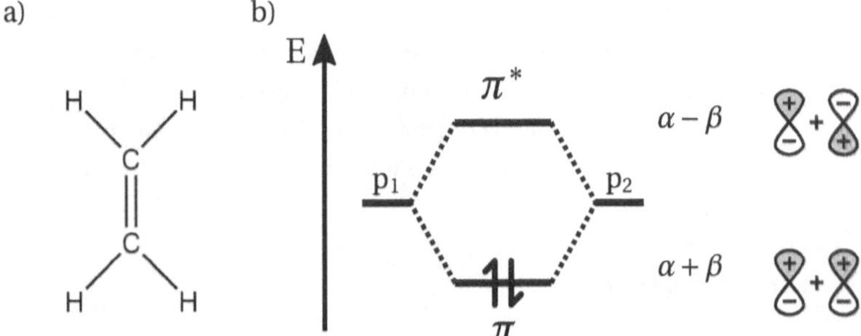

Figure 2.4.: a) Chemical structure of the ethene molecule. b) Energy levels and occupation of the π and π^* orbitals derived by the Hückel molecular orbital theory. A visualization of the p orbital linear combinations is shown on the right.

2.1.3. Van der Waals Bonding

As described in many standard textbooks on organic semiconductors, *e.g.* in [SW07] which served as a basis for this section, the formation of stable solid state aggregates out of non polar molecules is enabled by the *van der Waals* interaction. The charge distribution of a molecule is not rigid, but rather subject to temporal fluctuations, in return causing temporally fluctuating dipole moments. These dipole moments can then induce a dipole moment in a nearby molecule leading to an attraction between the two dipoles.

In the far field approximation, the electric field of a dipole $\vec{p}_1$ at a position $\vec{r}$ is given by

$$\vec{E} = \frac{1}{4\pi\epsilon_0} \frac{\left(3\left|\vec{p}_1\right|\vec{e}_r\cos\vartheta - \vec{p}_1\right)}{|\vec{r}|^3}, \tag{2.3}$$

with ϑ being the angle enclosed by $\vec{r}$ and $\vec{p}_1$. A molecule in the dipole field $\vec{E}$ at the position $\vec{r}$ will experience a displacement in its charge density inducing a dipole

$$\vec{p}_2 = \alpha\vec{E}(\vec{r}). \tag{2.4}$$

The *polarizability* α is measure of the deformability of the electron density, *i.e.* how easily electrons can be displaced in the atomic or molecular orbitals. The induced dipole $\vec{p}_2$ in the electric field of the initial dipole $\vec{p}_1$ has the potential energy

$$V = -\vec{p}_2\cdot\vec{E} = -\alpha\vec{E}\cdot\vec{E} = -\frac{\alpha\left|\vec{p}_1\right|^2}{(4\pi\epsilon_0)^2}\frac{\left(3\cos^2\vartheta + 1\right)}{|\vec{r}|^6}. \tag{2.5}$$

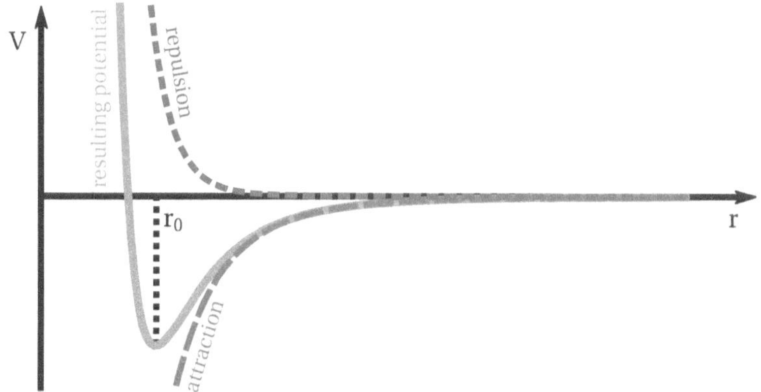

Figure 2.5.: Bonding potential in molecular solids described by the resulting potential (orange) of a repulsive (red) and an attractive component (green) as function of the inter molecular distance r. The distance r_0 marks the equilibrium distance. Adapted from [SW07].

In the context of small molecules, the polarizability is, of course, dependent on the individual atoms and their respective bonds comprising each molecule as well as on the relative respective orientation of the interacting molecules under consideration. For a qualitative description, equation (2.5) is simplified to $V_{vdW} = -A \cdot r^{-6}$ and describes the attractive potential between two non polar molecules as a function of the inter molecular distance r, called *van der Waals* interaction. The parameter A is unique for every molecule and highly anisotropic. The attractive potential as function of their intermolecular distance is shown by the green curve in figure 2.5. In close proximity to each other, the molecules experience a strong repulsive potential resulting from Coulomb repulsion of the nuclei, which, in contact, are less screened and the Pauli principle, *i.e.* the prohibition of two electrons with identical set of quantum numbers to occupy the same state. This repulsion, red curve in figure 2.5, together with the attractive van der Waals potential leads to the resulting potential with a bonding equilibrium position r_0, as shown in figure 2.5. The repulsive potential is frequently assumend as a reciprocal power law $\propto r^{-m}$. For $m = 12$ the resulting potential is known as the *Lennard-Jones potential*:

$$V = \frac{B}{r^{12}} - \frac{A}{r^6}. \tag{2.6}$$

To take the quantum mechanical nature of the repulsive potential into account, Max Born and Joseph E. Meyer instead used an exponential dependency for the repulsive potential [BM32]. With the *Born-Meyer potential* $V_{BM} = B\exp(-ar)$ the resulting potential, known as the *Buckingham potential* [BC47], is given as

$$V = Be^{-ar} - \frac{A}{r^6}.$$ (2.7)

Both potentials result in a strong repulsion for short inter molecular distances and a stable equilibrium distance, which is the requirement for the formation and stability of organic crystals and crystalline aggregates. [SW07]

2.2. Crystal Formation and Growth Processes

In [BW88] Joachim Bohm defines a crystalline state as a state "for which the majority of atomic building blocks (atoms, ions, molecules) are, over a more or less large area and long time period, arranged in a regular three dimensional periodic manner, leading to a long-range order and enabling an assignment to a lattice system".[1] This definition includes that a small amount of defects is part of a real crystal. However, they can be considered negligible, as long as their appearance is rare on length scales (and also timescales if one wants to include diffusion processes within the crystal) of the physical property that is being examined.

Crystals can be grown by various techniques and from various states of matter, *i.e.* liquid, gas and even solids. The following section will concentrate on crystal formation from a single kind of molecule out of the gaseous phase being of relevance for this work. For a complementary and more detailed discussion of the various growth mechanisms reference is made to [BW88], [KBB90] and [Mar10], from which the following sections 2.2.1 and 2.2.2 are compiled.

2.2.1. Crystal Formation: Nucleation from Supersaturated Vapor

Crystal growth can roughly be divided into three basic steps: Supersaturation, nucleation and growth of the nuclei into a macroscopic single crystal [Mar10]. The thermodynamic potential used for the corresponding description is the *Gibbs potential* or *free enthalpy G*. The equilibrium state of a system at given pressure P, temperature T and with the amount of substance n is found as a minimum of $G(P, T, n)$. The Gibbs potential is usually defined as

$$G = U + PV - TS$$ (2.8)

---

[1] Author's loose translation from German original.

with the internal energy U, volume V and entropy S. For a single substance it is valid to write $G = n\mu$ with μ being the chemical potential in the respective phase. If a substance has two phases α and β the coexistence curve in the corresponding phase diagram is described by the *Clausius-Clapeyron* equation:

$$\frac{\mathrm{d}P}{\mathrm{d}T} = \frac{1}{T}\frac{\Delta H}{\Delta V}. \tag{2.9}$$

In this equation ΔV is the change in volume upon the $\alpha \rightarrow \beta$ phase transition and ΔH is the change in enthalpy, in case of a solid to gas change, the enthalpy of sublimation. [BW88]

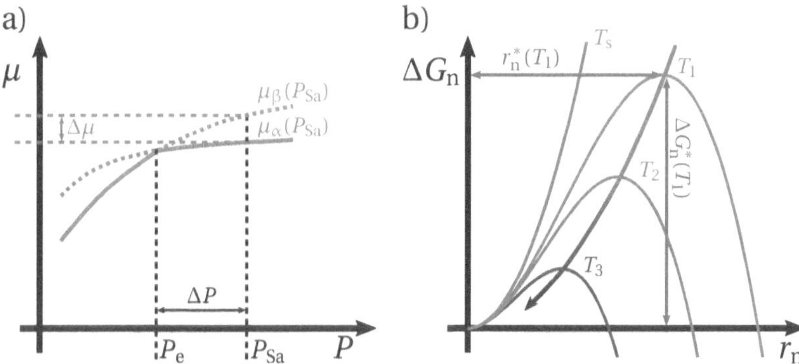

Figure 2.6.: a) Pressure dependence of chemical potentials of a substances condensed phase (α, blue) and gaseous phase (β, green) at temperature T. At the vapor pressure P_{e} both phases coexist in equilibrium, branching into their respective stable (solid line) and metastable (dashed line) configurations. For supersaturated β vapor at pressure P_{Sa} condensation to the α phase reduces the systems chemical potential by $\Delta\mu$. b) Change in system's Gibbs potential ΔG_{n} during nucleation as function of nuclei dimension r_{n} at condensation temperatures $T_1 > T_2 > T_3$. If the system is kept at sublimation temperature T_{s}, stable nuclei can not form, as the vapor and the solid phase are in equilibrium. Both adapted from [BW88].

If a substance is in its gaseous phase β at a constant temperature T, its free enthalpy and hence, its chemical potential will increase monotonously with increasing partial pressure. If a solid state phase α exists, this increase continues up to the equilibrium partial pressure P_{e} where the two phases coexist. This pressure is called vapor pressure. The qualitative course of the chemical potentials of both phases is shown exemplarily in figure 2.6 a). Solid lines indicate a stable phase, while a dashed line indicates that the respective phase is metastable in this chemical potential-pressure regime. Stabilizing the β phase above the vapor pressure leads to supersaturation of the gaseous phase. Condensation of the supersaturated vapor at P_{Sa} minimizes the energy of the system by $\Delta\mu = \mu_\alpha(P_{Sa}) - \mu_\beta(P_{Sa}) < 0$ per mol. This difference in chemical potential between the two phases is the driving force of the nucleation process and can be approximated by using the ideal gas law with the ideal gas constant R as

$$\Delta\mu(P_{Sa}) = -RT\ln\frac{P_{Sa}}{P_e}. \tag{2.10}$$

Subliming a substance at a temperature T_s and condensing the respective vapor at a temperature $T_c < T_s$, equation (2.10) yields the maximal achievable difference in chemical potential of

$$\Delta\mu = -RT_c\ln\frac{P_{e,s}}{P_{e,c}} \tag{2.11}$$

where $P_{e,s}$ and $P_{e,c}$ are the respective vapor pressures at the sublimation and condensation temperature. Integrating the Clausius-Clapeyron equation (2.9) and using the ideal gas law for the volume change $\Delta V = nRT/P_e$ the logarithm of the vapor pressure ratio is

$$\ln\frac{P_{e,s}}{P_{e,c}} = \frac{\Delta h\,(T_s - T_c)}{RT_s T_c} \tag{2.12}$$

where Δh is the molar heat of transformation from the condensed to the gaseous phase.

The process of nucleation from a supersaturated vapor (amount of substance n, supersaturation pressure P_{Sa}, temperature T_c) can be described by analyzing changes in the Gibbs potential G of the system, under the assumption that the resulting nuclei are big enough to be assigned macroscopic thermodynamic stable quantities [Mar10]. Starting from supersaturated vapor, the system has the free enthalpy $G_1 = n_\beta\mu_\beta(P_{Sa}) = n_\beta\mu_\beta$. If a fraction of the vapor n_α undergoes a phase transformation to the condensed phase α, the free enthalpy of the system changes to $G_2 = (n_\beta - n_\alpha)\mu_\beta + G_n(n_\alpha)$ with $G_n(n_\alpha)$ being the free enthalpy of the newly formed α phase nuclei. The change in free enthalpy is then given by

$$\Delta G = G_2 - G_1 = G_n(n_\alpha) - n_\alpha\mu_\beta. \tag{2.13}$$

For simplicity's sake, a cubic geometry with linear dimension r_n for the nuclei can be assumed without affecting the description's result. Actually, this assumption can be justified for a mono molecular crystal with a simple cubic lattice. For other geometries the description becomes more complicated and the assumption of an isotropic surface tension cannot be maintained. The amount of substance can be rewritten by

$n_\alpha = r_\mathrm{n}^3/v_\alpha$, using the molar volume v_α. Considering the nuclei as a closed volume, the surface tension γ needs to be accounted for leading to a free enthalpy for the nuclei of

$$G_\mathrm{n}(r_\mathrm{n}) = \frac{r_\mathrm{n}^3}{v_\alpha}\mu_\alpha + 6r_\mathrm{n}^2\gamma. \tag{2.14}$$

Inserting (2.14) in (2.13) leads to the following expression for the change in free enthalpy due to nucleation as a function of the nuclei dimension r_n

$$\Delta G_\mathrm{n}(r_\mathrm{n}) = \frac{r_\mathrm{n}^3}{v_\alpha}(\mu_\alpha - \mu_\beta) + 6r_\mathrm{n}^2\gamma = \frac{r_\mathrm{n}^3}{v_\alpha}\Delta\mu + 6r_\mathrm{n}^2\gamma. \tag{2.15}$$

In figure 2.6 b) ΔG_n is plotted as a function of the linear nuclei dimension r_n for different temperatures $T_1 > T_2 > T_3$. The difference in chemical potential of the two phases $\Delta\mu$ is a negative quantity and can be calculated from (2.11). Taking a closer look at the temperature T_1 an increase of ΔG_n with increasing r_n is evident reaching a maximum at r_n^* as shown in figure 2.6 b). For larger r_n the difference in the Gibbs potential decreases rapidly and becomes negative. This means that for nucleation to be the system's energetically favorable configuration, the nuclei formation work G_n^* needs to be provided by thermal fluctuations. The length r_n^* is the critical dimension of a forming nuclei, as nuclei with $r_\mathrm{n} < r_\mathrm{n}^*$ are unstable and only nuclei with $r_\mathrm{n} > r_\mathrm{n}^*$ can achieve a stable configuration and are able to keep on growing into macroscopic crystals. A critical nuclei with $r_\mathrm{n} = r_\mathrm{n}^*$ forms a metastable configuration and can develop either way. The critical dimension r_n^* can be calculated as the position of the maximum of ΔG_n using $\partial\Delta G_\mathrm{n}/\partial r_\mathrm{n} = 0$. This leads to

$$r_\mathrm{n}^* = -4\frac{\gamma v_\alpha}{\Delta\mu} \tag{2.16a}$$

$$\Delta G_\mathrm{n}^* = \Delta G_\mathrm{n}(r_\mathrm{n}^*) = 32\frac{\gamma^3 v_\alpha^2}{\Delta\mu^2} \tag{2.16b}$$

for the critical size r_n^* and the nucleation work G_n^*.

The dependency of the energy needed for nuclei formation ΔG_n^* and of the respective nuclei size r_n^* on the condensation temperature can be approximated by equating (2.11) and (2.12) resulting in

$$\Delta\mu = -\Delta h\left(1 - \frac{T_\mathrm{c}}{T_\mathrm{s}}\right). \tag{2.17}$$

Therewith, expressions (2.16a) and (2.16b) giving the critical parameters read as

$$r_n^* = 4\frac{\gamma v_\alpha}{\Delta h(1 - T_c/T_s)} \tag{2.18a}$$

$$G_n^* = 32\frac{\gamma^3 v_\alpha^2}{\Delta h^2(1 - T_c/T_s)^2}. \tag{2.18b}$$

Their correlated temperature dependence is shown in figure 2.6 b). For $T_c = T_s$ the critical size tends towards 0 while the required energy diverges. No stable nuclei can be formed, as the system is in thermodynamic equilibrium at its vapor pressure. As the condensation temperature decreases, r_n^* and ΔG_n^* decrease as well, leading to the formation of smaller nuclei. As a finite amount of energy is needed to enable the formation of stable nuclei and hence, the increase in free enthalpy by ΔG_n, the formation probability is proportional to $\exp(-\Delta G_n/k_B T_c)$ with k_B being the Boltzmann constant [Ein10]. This means that the rate of nuclei formation can be estimated as

$$J = J_0 \exp\left(-\frac{\Delta G_n^*}{k_B T_c}\right) \tag{2.19}$$

with J_0 representing a material dependent constant, which has to be deduced individually using molecular kinetic theory [BW88] [KBB90]. Therefore, the formation rate as well as the dimensions of the resulting nuclei is sensitive to changes in the sublimation as well as the condensation temperature.

2.2.2. Crystal Growth

After the successful formation of a stable nucleus, the formed crystal grain will start growing, given suitable conditions, up to a macroscopic crystal. The shape of the crystal, *i.e.* the form and size of the respective crystal faces, is referred to as *crystal habit* and is primarily determined by the *growth pace* of the different crystal faces. The growth pace is defined as the distance a crystal facet grows along its surface normal per time interval. Faces with a slower growth speed will persist through the process and become larger on the expanse of the faster growing facets, which, *vice versa*, can be fully seized by a single dominant crystal facet.

In figure 2.7 a) the schematics of a growing nucleus with (100),($\bar{1}$00),(001),(101) and a ($\bar{1}$01) facet is shown at successive time steps during the growth process. The correlated growth paces v_{hkl} of the different crystal facets follow the relation $v_{100} > v_{001} > v_{101}$. The (001) facet front pushes along the [001] crystal direction, while the slower evolving (101) and ($\bar{1}$01) faces simultaneously, are dragged along increasing their surface areas and therby constricting the "driving" (001) surface as indicated by the purple arrows.

Eventually, the same mechanism leads to the disappearance of the two (h00) faces. Their narrowing is indicated by purple arrows as well.

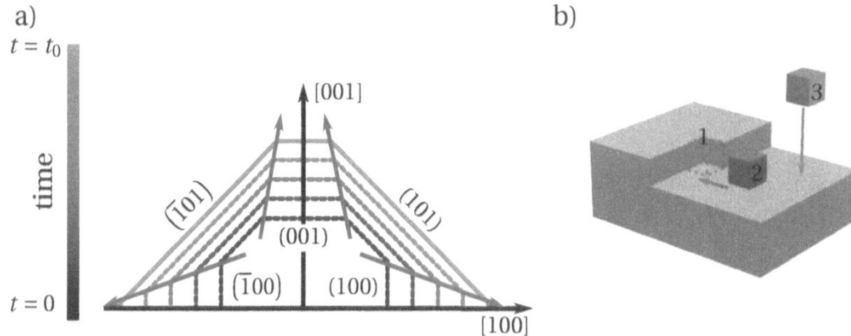

Figure 2.7.: a) Crystal facet growth at consecutive time steps. The slower growth pace of the ($h0l$) faces leads to a narrowing of the (001) facet and the full disappearance of the two (h00) surfaces. The utilized parameter in arbitrary length and time units are $v_{100} = 2$, $v_{001} = 1.5$ and $v_{101} = 0.6$. Constriction of the shrinking facets is indicated by purple arrows. b) Kossel crystal growth model: A particle in the gaseous phase (3) can adsorb on the crystal surface (2) and diffuse to a binding half-crystal position (1). This process successively completes the next layer on the crystal surface. Both adapted from [KBB90].

Modeling the growth process by the facets' progression at different growth speeds explains the crystal habit well, but the microscopic progresses underlying the facets' growth are not considered at all. Various different microscopic mechanisms are involved in the growth process and a full recitation would exceed the scope of this work, which is why only the basic concepts are discussed in the following.

A microscopic model to explain the observed growth on an already existing surface was developed independently from each other by Walther Kossel [Kos27] and Iwan Stranski [Str28] in the late 1920s, using the crystal growth of NaCl crystals as a reference system. Even though the model was initially deduced for the ionic binding based on the Coulomb interaction between ions, the general findings can be transferred to many other systems. Assuming that the crystal structure on the crystal surface does not differ significantly from the structure inside the crystal and hence, the interaction between two crystal constituents on the surface and inside the crystal on equivalent lattice sites is described by the same interaction potential, the principle of energy minimization does explain the crystal growth [Str28]. Starting with any kind of attractive interaction between two building blocks[2], a particle positioned close to another particle will be

---

[2] For the Coulomb interaction this is, of course, not necessarily the case and can lead to interesting growth phenomena. For example in the case of NaCl crystals, the expression of the (110) surface is suppressed [Kos27].

energetically favorable with respect to a free particle. Consider an atomically smooth surface, *i.e.* no roughness on height scales of the crystal's building blocks, which of course holds true for molecules as well, with terrace formed by an additional layer on top as shown in figure 2.7 b). The particle attached to the surface at position 2 to is in an energetically more favorable position than a free particle in the gaseous phase as exemplary shown at position 3. Position 1 is called *half-crystal position* and is the energetically most favorable position provided by a kink in the step edge. There, the simplified cubic building block has interaction neighbors on three of its six sides. Particles in the gaseous phase can be adsorbed on the crystal surface and diffuse towards the kink, filling up the energetically favorable half-crystal positions. After one row is completed, the next row forms until the layer is completed. The step edge usually is not generated from scratch, starting with a single particle on a smooth surface, growing into a row *etc.*, because desorption of particles from the surface due to the thermal energy of the system is a non neglectable process. Instead a number of particles form an incidental surface nucleus, meaning a two dimensional aggregate with epitaxial relation to the crystal structure of the underlying crystal facet. The process is comparable to the 3D nucleus formation described in the previous section 2.2.1 if the surface interaction between facet and nucleus is taken into account. Hence, the growth pace of the crystal facet is highly sensitive to the nucleation rate on the surface, which can be expressed similarly to equation (2.17). This means that the growth temperature as well as the supersaturation of the environment are important parameters determining the facet growth speed and hence the overall crystal growth. For the sake of completeness, it should be mentioned that in some cases the nucleation rate cannot be directly related to the observed growth speed, especially for low supersaturation. Other process however, such as the screw dislocation assisted growth mode, which shall not be explored here and is described in literature [BW88], can explain this mismatch. There, the terrace step edge on the crystal facet originates from a screw dislocation on the growing crystal surface.

Rarely one surface nucleus is formed and ceases its growth over the surface before the whole next nucleus is generated. Rather several surface nuclei grow simultaneously next to each other leading to an *island like growth* and on top of each other, forming a so called *vicinal* surface. An atomic force microscopy (AFM, see chapter 3.3.1) image of the surface of a pentacene single crystal is shown in figure 2.8 a). Terraces of (001) facets staple on top of each other, characterized by mono molecular steps at which additional molecules can adsorb, leading to lateral *step-flow growth*. The front of the propagating step edges forms an apparent macroscopic surface which is slightly tilted with respect to the physical terrace planes.

a)   b)

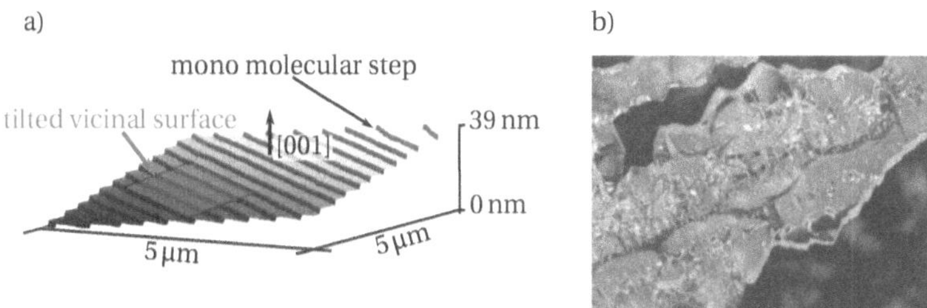

Figure 2.8.: a) Three dimensional (3D) representation of a pentacene single crystal surface recorded via AFM. The mono molecular steps appearing at the edges of the (001) terraces form a vicinal surface. b) Optical microscope image of a DBTTF single crystal with dendritic growth produced by horizontal vapor deposition under extreme supersaturation.

If the diffusive particle transport to the positions with the lowest energy on the crystal surface can not compete with the impinging particles, crystal surfaces can not complete their growth process and new material is added preferably to macroscopic crystal edges leading to a dendrite-like growth or, for more extreme conditions, even to a diffusion limited growth. Dendrite growth is not uncommon for organic crystals and, for example, has been observed during the growth of crystalline tetracene in liquid thin films [Voi+03]. As anothoer example, figure 2.8 b) shows a microscope image of a dibenzotetrathiafulvalene (DBTTF) crystal grown via horizontal vapor deposition (*c.f.* section 3.2.1) at exceeding supersaturation. The triangular shapes growing from the central crystal body and the rough surface are common for dendritic growth.

2.2.3. Phase Transition and Kinetics

Solid state structural phase transitions, *i.e.* the rearrangement of the crystal structure, is a quite common phenomena, also for organic molecular crystals. It is well established that fullerene C_{60} changes from a *face-centered cubic* structure at room temperature to a *simple cubic* crystal structure at about 250 K [Hei+91] and undergoes a second transition at 90 K where the rotation of the molecules is frozen-in [Dav+92]. Another example, out of many, is the rearrangement of the pyrene $C_{16}H_{10}$ dimer structure below 100 K [Kni+96][Fra+00]. It is quite obvious that the crystal structure and hence, the relative molecular position, influences the macroscopic properties of a crystal like its optical absorption or emission (*c.f.* section 2.3.3) as well as its electric conductivity.

Qualitatively, a phase transition can be described as follows [Fli10]: As discussed in section 2.2.1, at a pressure P and temperature T, the equilibrium configuration of a material is the phase of lowest Gibbs potential, which is equivalent to the chemical potential μ if the particle number is conserved within the system. The potential surface $\mu(\tilde{p}_1, ..., \tilde{p}_n)$ can have several minima for different sets of structural parameters $(\tilde{p}_1, ..., \tilde{p}_n)$. The $\tilde{p}_i$ have to be interpreted as general structural parameters and can represent any variable influencing the chemical potential like lattice distances and angles, atomic positions or molecular deformations. This means, the present phase is not necessarily the one with the global minimal energy, but could be a local minimum. These configurations are called *metastable* and can for example be converted to a lower energy configuration by adding thermal energy to overcome a potential barrier. Figure 2.9 sketches a chemical potential depending on a single generic structural parameter $\tilde{p}$. Here, position 1 defines a metastable potential minimum. Applying the energy $\Delta\mu$ to overcome the potential barrier the system can be transferred to the global potential minimum 2. The strive to minimize its energy can be interpreted as the driving force of the phase transition.

A full thermodynamic description and classification of solid state phase transitions is given by the *Landau theory*. As this is extensively discussed in literature, for example in [Cow80]. Apart from the qualitative discussion above it will not be discussed in this work. However, for a complementary discussion, a microscopic model of the phase transition kinetics will be briefly presented.

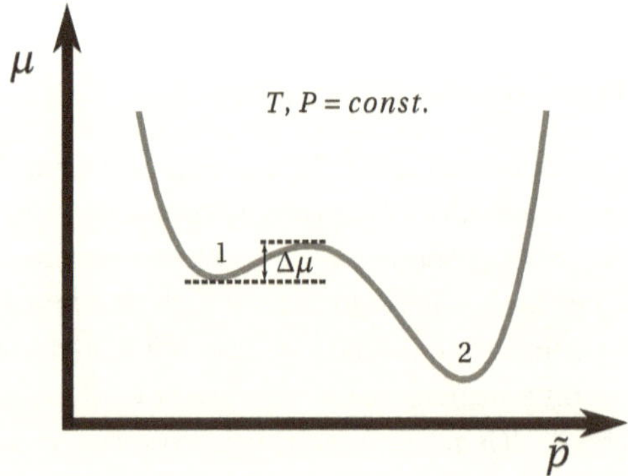

Figure 2.9.: Chemical potential as function of generic structural parameter $\tilde{p}$. The metastable configuration 1 can be converted into the stable configuration 2 by supplying the necessary energy $\Delta\mu$ to overcome the potential barrier.

2. Theoretical Background

Since the 1940s, the *Johnson-Mehl-Avrami-Kolmogorov*[3] *(JMAK)* model has been used as a phenomenological approach to describe the kinetics of isothermal phase transitions and can be summarized in the simple formula

$$V(t) = 1 - e^{-V_e(t)} \tag{2.20}$$

where $V(t)$ is the transformed volume fraction at time t, and $V_e(t)$ is the *extended Volume* at time t, *i.e.* the volume of the transformed phase ignoring overlap of its constituting, continuously growing grains [FT98]. Melvin Avrami derived equation (2.20) in the second of his famous paper series on "the kinetics of phase change" [Avr40] from probability considerations on the ratios between transformed, non transformed and extended volume.

Consider a very large, under idealized conditions, infinite volume in a metastable phase α and assume the stable phase β to form from growing β phase grains throughout the volume. The growth of the grains occurs by transformation of the α phase to the stable β phase along the boundary of the two phases at an isotropic growth rate G. The transforming grains originate from point-like grain nuclei which form at a rate $N(t)$ per volume due to thermal fluctuations. The volume of a single grain at time t which originated from a nuclei formed at time t_0 is given by $v(t, t_0)$. For the volume of all grains at time t, disregarding any grain overlap, hence the extended volume, reads [Avr39]

$$V_e(t) = \int_0^t v(t, t_0) N(t_0) \, dt_0. \tag{2.21}$$

If the growth is sufficiently benign, the grain volume can be expressed using a shape factor σ as [WBS97][Avr39]

$$v(t, t_0) = \sigma \left(\int_{t_0}^t G \, dt' \right)^3 = \sigma G^3 (t - t_0)^3 \tag{2.22}$$

assuming a constant growth speed. In equation (2.22) $G(t - t_0)$ is a length defining the extend of the growing grain at time t. For spherical grains $\sigma = 4/3\pi$. With equations (2.21), (2.22) and the assumption of a constant nucleation rate N the expression (2.20) becomes

$$V(t) = 1 - e^{-\frac{\sigma G^3 N}{4} t^4} \tag{2.23}$$

which is the well established "t to the power of 4" -law for a polyhedral, meaning three dimensional isotropic grain growth [Avr40]. Under similar assumptions the extended

---

[3]For a historical summary of the respective contribution of the four scientists, it is referred to [FT98] as here, only the work of Melvin Avrami will be discussed.

volume term for a two dimensional *plate like* growth and a one dimensional *needle like* growth can be derived from equations (16') and (16") in [Avr40] yielding

$$V_{e,2D} = \frac{\sigma G^2 N}{3} t^3 \tag{2.24}$$

for the two dimensional case and

$$V_{e,1D} = \frac{\sigma G N}{2} t^2 \tag{2.25}$$

for the needle like one dimensional growth.

As the assumptions are quite restrictive and rarely met in real world systems, advanced theoretical work has been conducted to extend the model to a variety of cases which are discussed to some extend in [FT98] and [WBS97].

2.3. Excitation of Molecular Aggregates

Light-matter interactions and the associated electronic transitions are of fundamental interest in solid state physics. They constitute the fundamental concepts to address a broad range of questions on electronic structure and interaction as well as yielding the basis for understanding, developing, and further optimization of optoelectronic devices, *e.g* light emitting diodes or photovoltaic cells.

This chapter is concerned with photo induced excitations, starting with an overview of the particularities of electronic transitions and excited state dynamics in single molecules. Subsequently, excitations in crystalline solids are discussed followed by a review on an important type of excited interface state, the *charge transfer exciton*.

2.3.1. Electronic Transitions in Single Molecules

The transition kinetics between two electronic states induced by either the absorption or the emission of a photon can be described by a rate model using the Einstein coefficients, if a not "too coherent" excitation is considered. Otherwise, if the energy gap is in resonance with the frequency of a coherently oscillating light field, the transition rates are no longer time-independent but undergo oscillations [Ger10]. This phenomenon is known as *Rabi oscillations* and has been observed for molecular two level systems [Ger+09]. However, it is out of scope of the present work.

In a two level system with ground state $|1\rangle$ and excited state $|2\rangle$, the Einstein coefficients A and B are defined such that

- the absorption and stimulated emission rates are given by $B_{12}u(\omega)=B_{21}u(\omega)$ with the spectral energy density $u(\omega)$

- the spontaneous emission rate is given as A_{21} [TL10].

First introduced by Einstein as a hypothetical construct to derive the Planck black body radiation formula [Ein17], these transition rates can be derived from time dependent perturbation theory. For this purpose a system comprising a single molecule and a present light field is described by the Hamiltonian

$$\hat{H} = \underbrace{\hat{H}_{\mathrm{mol}} + \hat{H}_{\mathrm{L}}}_{=\hat{H}_0} + \hat{H}^{\mathrm{S}} \tag{2.26}$$

where the ground state Hamiltonian $\hat{H}_0$ comprises the molecular part $\hat{H}_{\mathrm{mol}}$ containing the electronic and nuclear interactions as well as the energy of the light field $\hat{H}_{\mathrm{L}}$. The interaction between the molecule and the light field is included as perturbation $\hat{H}^{\mathrm{S}}$. As the ground state Hamiltonian can be described as a sum of the separate parts of the system, the resulting wave function can be written as a product

$$\Psi = \psi_{\mathrm{Mol}}\phi_{\mathrm{L}} \tag{2.27}$$

with the molecular wave function ψ_{Mol} and the light field wave function ϕ_{L}. The dipole approximation, valid for a single electron and the light's wavelength being large compared to the extend of the electron's wave function, is not necessarily applicable to larger molecules. The light field is described by the vector potential operator $\hat{A}$. The interaction with a single charge carrier l at position $\vec{r}_l$ of mass m_l, charge q_l, and the respective momentum operator $\hat{p}_l$ is given by

$$\hat{H}^{S'} = \frac{q_l}{m_l}\hat{A}(\vec{r}_l)\cdot\hat{p}_l + \frac{q_l^2}{2m_l}\underbrace{\hat{A}^2(\vec{r}_l)}_{\approx 0} \tag{2.28}$$

where $\hat{A}(\vec{r}_l)$ is a function of the charge carrier's position. For a light field of sufficiently small amplitude, the quadratic term in (2.28) can be neglected. For a molecule consisting of M atoms with their nuclei at $\vec{R}_K$ and atomic number Z_K and N electrons at locations $\vec{r}_j$ the light matter interaction is then described by

$$\hat{H}^{S} = \underbrace{\sum_{j=1}^{N}\left(-\frac{e}{m_0}\hat{A}(\vec{r}_j)\cdot\hat{p}_j\right)}_{\text{electronic part } \hat{H}_{\mathrm{e}}^{S}} + \underbrace{\sum_{K=1}^{M}\left(\frac{eZ_K}{M_K}\hat{A}(\vec{R}_K)\cdot\hat{P}_K\right)}_{\text{nuclear part } \hat{H}_{\mathrm{n}}^{S}}. \tag{2.29}$$

The energy of the light field is determined by the number of photons at each energy. The interaction Hamiltonian describing the absorption or emission of a photon by the molecule, acts on ϕ_L by removing or adding a photon of the respective energy from or to the light field. Focusing on photons of sufficient energy, the electronic part of the Hamiltonian becomes the dominating part of the interaction. Under this assumption, the generation (or annihilation) of a photon with energy $E_{ph} = \hbar\omega$ by the interaction Hamiltonian is accompanied by the simultaneous transition of the molecule's electronic wave function ψ^e_{Mol} from a state of higher energy (E_h) to a state of lower energy (E_l) (or *vice versa*), thereby conserving the system's overall energy by obeying the relation

$$E_{ph} = \hbar\omega = \Delta E = E_h - E_l. \tag{2.30}$$

As a basic result from perturbation theory, often referred to as *Fermi's golden rule*, the matrix elements of the perturbation operator are directly linked to the transition rate between an initial state $|i\rangle$ and a final state $|f\rangle$ as

$$k_{if} = \frac{2\pi}{\hbar} \sum_{\Gamma} \hspace{-1.1em}\int \; \left| \langle f | \hat{H}^S | i \rangle \right|^2 \delta\left(E_f - E_i\right). \tag{2.31}$$

The initial and final state refer to the whole quantum system, *i.e.* the light field and the molecule, including the electronic, the nucleic and the spin part of the molecular wave function. Thus, the energies in the delta-distribution in (2.31) relate to the overall energy of the system before and after photon absorption or emission, meaning the transition rate only differs from zero when $E_i = E_f$. Hence, the delta-distribution assures the conservation of energy given by (2.30). The sum, respectively the integral, ensures counting all equivalent final states Γ, which can either be discrete (sum) or lie in a (quasi-)continuum (integral). This, of course, includes again the overall states of the quantum system. For example, an emitted photon is constrained in its wavelength by the conservation of energy and the chosen polarization in the matrix element, but the direction of its wave vector $\vec{k}$ is not determined by any restraints and therefore defines a quasi continuum for the emitted photon, hence for the final state.

Using the two level system notation the Einstein coefficients and the transition probabilities per second for absorbing or emitting a photon of frequency ω and hence, energy $\hbar\omega = |E_i - E_f|$, and of polarization $\vec{e}_\lambda$ in the solid angle $d\Omega$ are related in the following way:

$$\text{spontaneous emission:} \quad k_{21} = A_{21} d\Omega \tag{2.32a}$$

$$\text{absorption:} \quad k_{12} = \rho(\omega, d\Omega) B_{12} d\Omega \tag{2.32b}$$

where $\rho(\omega, d\Omega)$ is the respective spectral energy density, implying the expected dependency of the transition rate for photon absorption on the intensity of the incident light.

A full derivation of the transitions rates from equations (2.26) - (2.29) and (2.31) can be found in [HW03], which also was the foundation of the introduction above. Furthermore, two important results from the discussion in [HW03] shall be noticed. At first, the relation of the transition rates for stimulated emission, hence absorption, and spontaneous emission with frequency ω, usually derived for atoms in the dipole approximation [Par07],

$$A_{21} \propto \omega^3 B_{12} \tag{2.33}$$

is also valid for the more general case including large molecules [HW03]. If the dipole approximation is valid, *i.e.* if the evolution from the initial state $|i\rangle$ to the final state $|f\rangle$ is essentially resulting from the perturbation of a single electron by the light field, and the molecule is small compared to the wavelength of the absorbed or emitted photon, the following simple relations between the transition dipole moment $\vec{M}_{if} = \langle f|\hat{\mu}|i\rangle$, with the dipole operator $\hat{\mu} = e\hat{r}$ and the Einstein coefficients can be derived:

$$A_{fi} = \frac{\omega_{if}^3}{8\pi^2 \hbar \epsilon_0 c^3} \left| \vec{e}_\lambda \cdot \vec{M}_{if} \right|^2 \tag{2.34a}$$

$$B_{if} = \frac{\pi}{\hbar^2 \epsilon_0} \left| \vec{e}_\lambda \cdot \vec{M}_{if} \right|^2 \tag{2.34b}$$

In equation (2.34a) and (2.34b) $\vec{e}_\lambda$ is the polarization vector of the absorbed or emitted photon. A more detailed discussion of the derivation of the transition rates above taken from [HW03] is given in appendix A.

So far, only the electronic part of the molecular wave function has been considered. For a full picture of the photon absorption and emission by molecules the nuclei as well as the spin multiplicity has to be considered leading to many different effects determining the dynamics upon photo excitation. These processes are, *inter alia*, described in the standard textbooks [KB15], [HW03] and [SW07] on which the following description is based if not stated otherwise. As the electrons are much lighter than the atomic nuclei, it is commonly assumed that the electron movement and the nuclei movement proceed independently. This implies that the molecular wave function can be expressed as the product of the electron wave function Ψ_{el}, the wave function of the atomic nuclei Ψ_{nuc} and the spin wave function Ψ_{spin}

$$\Psi_{mol} = \Psi_{el} \Psi_{nuc} \Psi_{spin}. \tag{2.35}$$

The rates of photon absorption and emission are determined by the squared transition dipole moment. For a molecule in the initial state $|\Psi_{\mathrm{mol},i}\rangle$ evolving to the final state $|\Psi_{\mathrm{mol},f}\rangle$ expressed by (2.35) this yields

$$
\begin{aligned}
&\left|\langle\Psi_{\mathrm{mol},f}|\,\vec{e}_\lambda\hat{\mu}\,|\Psi_{\mathrm{mol},i}\rangle\right|^2 \\
&= \left|\langle\Psi_{\mathrm{el},f}\Psi_{\mathrm{nuc},f}\Psi_{\mathrm{spin},f}|\,\vec{e}_\lambda\hat{\mu}\,|\Psi_{\mathrm{el},i}\Psi_{\mathrm{nuc},i}\Psi_{\mathrm{spin},i}\rangle\right|^2 \\
&= \underbrace{\left|\langle\Psi_{\mathrm{el},f}|\,\vec{e}_\lambda\hat{\mu}\,|\Psi_{\mathrm{el},i}\rangle\right|^2}_{\text{electronic transition }\vec{e}_\lambda\vec{M}_{\mathrm{el},if}}\left|\langle\Psi_{\mathrm{nuc},f}|\Psi_{\mathrm{nuc},i}\rangle\right|^2\left|\langle\Psi_{\mathrm{spin},f}|\Psi_{\mathrm{spin},i}\rangle\right|^2 \\
&= \left|\vec{e}_\lambda\vec{M}_{\mathrm{el},if}\right|^2\left|\langle\Psi_{\mathrm{nuc},f}|\Psi_{\mathrm{nuc},i}\rangle\right|^2\left|\langle\Psi_{\mathrm{spin},f}|\Psi_{\mathrm{spin},i}\rangle\right|^2
\end{aligned}
\tag{2.36}
$$

under the assumption that the electronic dipole operator only affects the electronic part of the wave function.

As it becomes evident, the transition dipole moment (2.36) comprises three distinct parts, each contributing to the overall transition rate. The last factor of the product, containing the spin states of the initial and final state, implies that an electronic transition is only allowed between electronic states of equal multiplicity. However, this is not strictly true for all transitions as will be discussed below. The factor containing the electronic states also includes the dipole operator $\hat{\mu} = e\hat{r}$ which is of odd parity under spatial inversion. This leads to the well-known selection rule that requests the final and initial electronic state to be of unlike parity, as otherwise the integral in (2.36) is zero and therefore, prohibiting a transition. The nucleic wave function contains the motion of the individual atomic nuclei, and hence the molecular vibrations and rotations. As the rotations are only observed in high resolution experiments on dilute gases, or *e.g.* He droplets, and are fully absent in solid states, they will not be discussed further. The vibronic part of the wave function has important implications on the spectra of electronic transitions and shall be discussed in some more details.

Considering an archetypical molecule consisting of only two atoms, the potential along the inter molecular axis is composed of an attractive part, caused by the binding molecular orbitals and, in close proximity, of a repulsive part rooted in the Pauli repulsion of the fully occupied lower atomic orbitals and the Coulomb repulsion of the atomic nuclei. Such potential curves are shown in figure 2.10 as a function of a general nuclear coordinate $\tilde{q}$. At first, the absorption process, as depicted in figure 2.10 on the left, shall be examined closer. The nuclear potential of the electronic ground state Σ has its equilibrium potential at $\tilde{q}_0$. Around this position the potential can be approximated as harmonic and hence, the molecular vibration can be described by an harmonic oscillator ansatz. The first excited state Σ' in organic molecules is usually reached by the promotion of a HOMO electron to the LUMO level, *i.e.*, from a bonding

π orbital to an anti-bonding π^* orbital. As the binding σ orbitals are not effected by this process, the curvature of the potential usually does not change significantly. However due to the anti-bonding character of the now occupied π^* orbital the equilibrium distance of the nuclei increases resulting in an equilibrium position of the Σ' potential at $\tilde{q}_0' > \tilde{q}_0$. As discussed before, the nucleic motion is much slower than the motion of the electrons. This means that after the "sudden" change of the energetic landscape due to an electronic transitions, the nuclei will follow and relax in the new potential. This model is named *Franck-Condon-principle* after its initiator James Franck [FD26] and developer Edward Condon [Con26][Con28]. In this model, electronic transitions proceed as vertical, as shown in the diagrams of figure 2.10. According to (2.36), the matrix element representing the transition probability for a transition from an initial state $|i\rangle = |n, v\rangle$ (electronic state n, vibronic level v) to a finial state $|f\rangle = |m, \xi\rangle$ is given by

$$k_{if} = \left|\langle f \,|\, \vec{e}_\lambda \hat{\mu} \,|\, i \rangle\right|^2 = \left|\langle m, \xi \,|\, \vec{e}_\lambda \hat{\mu} \,|\, n, v \rangle\right|^2 = \left|\vec{e}_\lambda \vec{M}_{\text{el}, nm}\right|^2 \underbrace{\left|\langle \xi | v \rangle\right|^2}_{\text{Franck-Condon factor}}. \tag{2.37}$$

Equation (2.37) shows that the rate of an electronic transition is directly proportional to the squared overlap integral of the concomitant vibrational wave functions which is called *Franck-Condon factor.*

For the photon absorption process consider a molecule in its electronic ground state Σ. As for organic molecules the vibrational energies are in the order of $100\,\text{meV}$ to $200\,\text{meV}$, it is commonly assumed that the only occupied vibrational state v at room temperature is the vibronic ground state[4]. Typical energies for the C–C and C=C stretching modes are $150\,\text{meV}$ and $200\,\text{meV}$, respectively [KB15]. The allowed transitions are determined by the energy of the vibrational mode and their intensities are mediated by the corresponding Franck-Condon factors $|\langle v'|0\rangle|^2$ as indicated in the absorption spectra in figure 2.10 (center). The decadic extinction coefficient ϵ (*c.f.* section 3.4.1) is linked to the Einstein coefficient by $\epsilon \propto \omega B_{12}$ [HW03]. For a direct link to the matrix element usually ϵ/E is plotted (*c.f. section 3.4.1*). For most molecules, the excited vibrational states quickly (in about $1\,\text{ps}$ [KB15]) relax to the ground state due to thermal exchange with the environment and the photon emission process from the excited state Σ' then starts with $v' = 0$. The allowed transitions are again given by the energy of the vibronic sublevels while the transition rate is again determined by the Franck-Condon factor $|\langle v|0\rangle|^2$. The luminescence intensity is directly proportional to the transition rate, *viz.* the Einstein coefficient. A relation between the absorption and emission rate is given by equation (2.33) as $A_{21} \propto \omega^3 B_{12}$. Scaling the emission spec-

---

[4]This common simplification fails if low energy vibrational modes, e.g. torsion modes, are considered, but for exemplification it will be proceeded with the molecule being in its vibrational ground state.

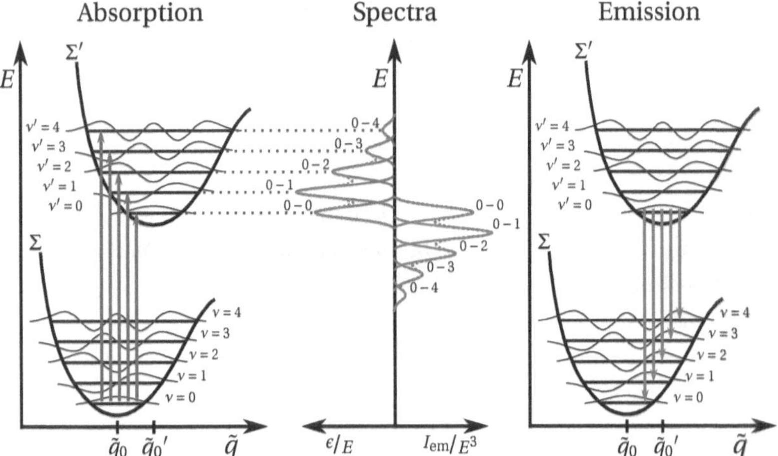

Figure 2.10.: Photon absorption (left) and emission (right) in the Franck-Condon model. The intensity of the spectral lines (center) is determined by the overlap integral of the vibronic wave functions indicated at the respective vibrational sublevels (horizontal lines in the potential curves). A weakening of the bonding upon transition from the electronic ground state Σ to the excited state Σ' leads to a shift of the equilibrium position. Adapted from [Ste15].

trum $I_{em}(E)$ as I_{em}/E^3 (*c.f. section 3.4.1*) reveals a mirror symmetry between absorption and emission spectra as a result of the Franck-Condon principle and can be found between absorption and emission spectra of various molecules [AGR06].

The above model assumes a displaced harmonic oscillator between the ground state Σ and the excited state Σ'. As can be seen, the Franck-Condon factor is mainly determined by the displacement of the equilibrium position $\Delta\tilde{q} = \tilde{q}_0' - \tilde{q}_0$ and the constant $k = \mu_r\omega_v^2$ of the oscillator with its reduced mass μ_r and vibrational frequency ω_v. From these quantities a dimensionless parameter

$$S = \frac{1}{2}\frac{\mu_r\omega_v}{\hbar}\Delta\tilde{q}^2 = \frac{^{1\!/_2}k\Delta\tilde{q}^2}{\hbar\omega_v}, \tag{2.38}$$

sometimes called the *Huang-Rhys parameter* or *factor*, is defined[Par07] [Jon+15]. It describes the coupling strength between the electronic transition and the vibronic modes and is a measure for the "displacement energy per vibrational quanta" as shown in (2.38). By means of the Huang-Rhys parameter the Franck-Condon factor for two vibrational levels m,n can be expressed as [Kei65b][Kei65a]

$$|\langle m|n\rangle|^2 = \begin{cases} e^{-S}S^{n-m}\left(\frac{m!}{n!}\right)\left(L_m^{n-m}(S)\right)^2, & \text{for } n \geq m \\ e^{-S}S^{m-n}\left(\frac{n!}{m!}\right)\left(L_n^{m-n}(S)\right)^2, & \text{for } n \leq m \end{cases} \tag{2.39}$$

with the associated Laguerre polynomials $L_n^{m-n}(x)$. At room temperature, *i.e.* $n = 0$, (2.39) yields

$$|\langle m|0\rangle|^2 = e^{-S}S^m\left(\frac{0!}{m!}\right)\left(\underbrace{L_0^{m0}(S)}_{=1}\right)^2 = \frac{e^{-S}}{m!}S^m, \tag{2.40}$$

which corresponds to a Poisson distribution of the transition intensities with mean value S.

Most molecules consist of $N > 2$ atoms, which leads to $Z = 3N-6$ vibrational modes[5]. If the vibrations are non interacting, which is usually is the case, their wave function can be expressed as a product of decoupled harmonic oscillators and the Franck-Condon spectrum for transitions to the respective ground state is given as [Par07]

$$F(E) = \sum_{j=1}^{Z}\sum_{v_j=0}^{\infty}\delta\left(\Delta E_{v_j,0} - E\right)\prod_{k=1}^{Z}|\langle\psi_{k,v_k}|0_k\rangle|^2. \tag{2.41}$$

Here, $|\Psi_i,v_i\rangle$ is the oscillator wave function of the i-th vibrational mode with quantum number v_i, while $|0_i\rangle$ is the respective ground state. The delta function marks the al-

---

[5]For linear molecules the number of modes is reduced by one

lowed transitions at $\Delta E_{v_i,0}$, *i.e.* the energy differences between the vibrational ground state of the initial electronic state and the final state in the vibrational state v_i of the i-th mode in the final electronic state.

Besides vibronic transitions, *i.e.* electronic transitions accompanied by changes in the vibrational states, other processes determine the molecule's photo dynamics. A convenient way to display an overview of these processes is provided by a *Jabłoński diagram* [Jab33] as shown in figure 2.11. It displays the singlet ground state S_0 and the first excited states of equal multiplicity S_n, as well as the first two triplet states together with their vibrational levels. For simplification nuclear potentials are not displayed.

The diagram implies that the triplet states are lower in energy than the excited singlet states and a phenomenological explanation originates from in the Pauli exclusion principle, leading to an on average larger spatial separation of the two electrons, and *vice versa*, to a lower Coulomb repulsion and thus, a lower overall energy [BBF01b]. Although this is often recited as an explanation and extension of Hund's rule, potential influences of the nuclei, which can have a significant impact on the stabilization of the triplet state as well, are disregarded [Rio07].

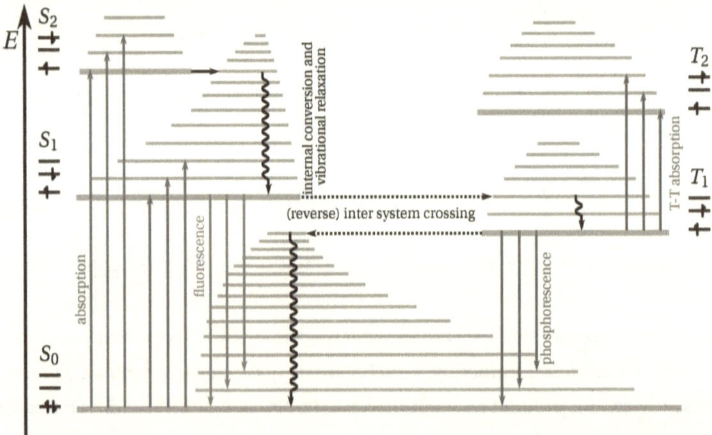

Figure 2.11.: Jabłónski diagram showing the various transitions between the different electronic and vibrational states of a molecule. Radiative transitions are depicted by different colors while non radiative transitions are labeled black. Adapted from [HW03].

After photon absorption, usually from the singlet ground state S_0 to a higher singlet state, the subsequent processes can be coarsely divided into radiative, if a photon

Process	Transition	Rate $k\ [\text{s}^{-1}]$	Reference
Fluorescence	$S_1 \rightarrow S_0$	10^9 $10^6\text{–}10^9$	[MS97], [Kas50] [SW07]
Phosphorescence	$T_1 \rightarrow S_0$ organic organometallic	 $10^{-2} \leq k << 10^6$ 10^6	 [SW07] [BBF01b] [KB15]
internal conversion	$S_n \rightarrow S_{n-1}$ $T_n \rightarrow T_{n-1}$	$10^7\text{–}10^{12}$	[KB15] [MS97]
vibrational relaxation	collisions phonon modes	10^{14} 10^{12}	[BBF01b] [MS97]
inter system crossing	$S_1 \rightarrow T_1$ organic organometallic	 $10^6\text{–}10^9$ 10^{12}	[KB15]

Table 2.1.: Representative rate constants of various excited state relaxation processes shown in figure 2.11.

absorption or emission is included, or non radiative, if no photon is involved. As described before, excited vibrational states quickly relax to the vibrational ground state of the respective electronic state due to thermal exchange with the environment [MS97]. This processes is called *vibrational relaxation*. An empirical law formulated in 1950 by Michael Kasha, and hence known as *Kasha's rule* states "The emitting electronic level of a given multiplicity is the lowest excited level of that multiplicity." [Kas50]. He estimates that the relaxation from higher states of the same multiplicity is roughly 10^4 times faster than the competing radiative transition. This means that fluorescence is usually the consequence of a $S_1 \rightarrow S_0$ transition [Kas50].

Kasha's phenomenological law is a consequence of the *inverse energy gap law*, which can be derived using perturbation theory transcending the Born-Oppenheimer approximation. A high energy state of any multiplicity Σ_n couples to a vibronic mode and is perturbed by the nuclear motion. A displacement in the vibronic ground state of Σ_n than leads to an evolution into an isoenergetic vibronic state of the energetically lower lying state Σ_{n-1}. This is known as *internal conversion*. From Fermi's golden rule (2.31) together with a linear approximation of the nuclear kinetic energy acting as perturbation operator a transition rate can be derived [Sie67]:

$$k_{\text{ic}} = \frac{2\pi}{\hbar} \rho J^2 F. \tag{2.42}$$

Here ρ denotes the density of states for the final state and J is the coupling of the electronic and the spin part of the molecular wave function to the displacement and the triggering vibronic mode, called the *promoting mode*. The vibrational mode populated after the transition from the initial electronic state Σ_n to the final electronic state Σ_{n-1} is called *accepting mode* and is labeled with k in the following. The promoting and accepting mode usually differ, as only the vibrational state with the lowest quantum number of all vibrational levels of Σ_{n-1} isoenergetic to the promoting mode's ground state contributes to the depopulation of Σ_n. The Franck-Condon factor F referrers to the overlap integral of the accepting mode's vibrational state in the electronic final state Σ_{n-1}, $|\nu_{\Sigma_{n-1},k}\rangle$, and the mode's ground state in the initial electronic state Σ_n, $|0_{\Sigma_n,k}\rangle$. It can be shown [Sie67] [KB15] that

$$F = \left|\langle \nu_{\Sigma_{n-1},k} | 0_{\Sigma_n,k} \rangle\right|^2 \propto \exp\left(-\gamma\frac{\Delta E}{\hbar\omega_k}\right) \tag{2.43}$$

where ΔE is the energy gap between the ground states of Σ_n and Σ_{n-1} and ω_k the accepting mode's frequency. The molecular geometry is included in the factor γ. As J is independent of the energy gap ΔE, only the Franck-Condon factor depends on the energy separation of the neighboring electronic states provoking energetically close states to undergo a fast non radiative conversion to lower lying states. The relation between equation (2.43) and (2.42) is know as the *inverse energy gap law*. The energetic difference between higher lying singlet states is usually much closer than the energy gap between S_0 and S_1, which explains the consistent experimental confirmation of Kasha's law. The same mechanism leads to *inter system crossing* where two states of different multiplicity are transferred into each other. The rate is much smaller as J includes the overlap between the spin wave functions and hence, starts from the lowest excited state. A transition between states of different multiplicity is only possible by spin orbit coupling, which will be described below. The reverse processes either leads to a lower lying singlet state or constitutes a thermally activated process from the T_1 to the S_1 state.

In general, transitions between pure states of different spin-multiplicity are forbidden, as the states' orthogonality reduces the overlap integral in equation (2.36) to zero. However, the coupling of an electron spin to its orbital momentum can lead to a perturbation of the pure states known as *spin-orbit interaction*. According to perturbation theory, this leads to higher energy states of different multiplicity admixed to the stationary state. In case of *phosphorescence, i.e.* the radiative transition from the T_1 state to the S_0 state, this means that the T_1 state is a linear combination of the pure triplet state and higher lying singlet states S_n. *Vice versa*, excited triplet states T_n also contribute to the singlet ground state S_0. The strength of these contributions depends on the matrix element of the spin-orbit interaction $\hat{H}_{SO}$, namely $\langle S_n | \hat{H}_{SO} | T_1 \rangle$ (singlets to triplet) and $\langle T_n|\hat{H}_{SO}|S_0\rangle$ (triplets to singlet), and inversely on the respective energy gaps $(E_{S_n} - E_{T_1})^{-1}$ and $(E_{T_n} - E_{S_0})^{-1}$. As the energetic spacing between the T_n-states and S_0 is, in general, much higher than the spacing of the excited singlet states and T_1, the admixture of singlet to the triplet states is the driving force of phosphorescence. In case of organic molecules, the time scale of a phosphorescent transition amounts to 1 ms to 1 second rendering fluorescence a much faster process (compare table 2.1). In single atoms, the strength of the spin-orbit interaction is directly proportional to the fourth power of the atomic number Z^4 [Fli18] and a similar expression can be found for molecules referring to the comprising atoms [KB15]. This means, adding heavy atoms to the molecular structure can dramatically enhance the inter system crossing and phosphorescence rate which led to the concept of phosphorescent OLEDs of high quantum efficiency [Bal+98].

An overview of the timescales of the various processes discussed in relation to figure 2.11 is given in table 2.1.

2.3.2. Dimers and Excimers

After discussing the excited states of a single molecule and their dynamics, inter molecular interaction between two molecules will be the topic of this section. Consider two identical monomers m_1 and m_2 with their respective ground states $|m_i\rangle$, $i = 1, 2$ with energy E_0, excited states $|m_i^*\rangle$ with energy E^* and an electronic transition dipole moment $\vec{M}_i$. In this simple model, the system can be described according to [SW07] and [KRE65] in the following way:

Without any interaction between the electronic systems, the eigenstates of the system's ground and excited states $|d_g\rangle$ and $|d^*\rangle$ are given as

$$|d_g\rangle = |m_1 m_2\rangle \tag{2.44a}$$

$$|d_\pm^*\rangle = \frac{1}{\sqrt{2}}\left(|m_1^* m_2\rangle \pm |m_1 m_2^*\rangle\right). \tag{2.44b}$$

As the molecules are identical their respective excitation probabilities are equal and hence, the excited state has to be constructed by a linear combination of both states. The molecular Hamiltonians $\hat{H}_i$ are independent. This leads to an energy of $2E_0$ for the ground state and $E_0 + E^*$ for the degenerated excited states $|d_\pm^*\rangle$.

Introducing a perturbation potential V_{12} describing the electronic interaction of the two molecules, the overall Hamiltonian of the dimer configuration is given by

$$\hat{H} = \hat{H}_1 + \hat{H}_2 + \hat{V}_{12}. \tag{2.45}$$

The states introduced in (2.44a) and (2.44b) are eigenstates of the system, but the degeneracy of the excited states $d_{\pm}^*$ is lifted by the perturbation. Calculating the energies of the dimer ground and excited state using the Hamiltonian (2.43) yields

$$\langle d_{\mathrm{g}} | \hat{H} | d_{\mathrm{g}} \rangle = 2E_0 + \underbrace{\langle d_{\mathrm{g}} | \hat{V}_{12} | d_{\mathrm{g}} \rangle}_{=D} \tag{2.46a}$$

$$\langle d_{\pm}^* | \hat{H} | d_{\pm}^* \rangle = E_0 + E^* + \underbrace{\langle m_1^* m_2 | \hat{V}_{12} | m_1^* m_2 \rangle}_{=D'} \pm \underbrace{\langle m_1^* m_2 | \hat{V}_{12} | m_1 m_2^* \rangle}_{=E_{12}}. \tag{2.46b}$$

The terms labeled D and D' in the above energies can be interpreted as van der Waals interactions between the two molecules in their ground state or between an excited and a non-excited molecule, respectively. For a stable dimer configuration, these contributions lead to an overall lowering of the respective states' energies, as indicated in figure 2.12 a). The term E_{12} describes the resonance interaction between the two excited states due to energy exchange and determines the energy splitting $\Delta E = 2E_{12}$ between the two states $|d_+\rangle$ and $|d_-\rangle$. From now on, the energetically higher and lower lying excited state of the two states $|d_{\pm}\rangle$ will be referred to as $|u^*\rangle$ and $|l^*\rangle$, respectively.

The Coulomb interaction between the molecules' electrons can be approximated by a multipole expansion and, if strong dipole transitions are allowed, can be simplified as dipole-dipole interactions [MK64]. For these, the resonance energy is expressed as

$$E_{12} = \frac{1}{4\pi\epsilon_0}\left(\frac{\vec{M}_1 \cdot \vec{M}_2}{r_{12}^3} - \frac{3\left(\vec{M}_1 \vec{r}_{12}\right)\left(\vec{M}_2 \vec{r}_{12}\right)}{r_{12}^5} \right) \tag{2.47}$$

with $\vec{r}_{12}$ being the position vector of the transition dipole moment of molecule 2 with respect to molecule 1 [KRE65]. The transition dipole moments for the dimer are given by [SW07]

$$\begin{aligned}
\vec{M}_{\mathrm{D}}^{\pm} &= \langle d_{\mathrm{g}} | e\hat{r} | d_{\pm}^* \rangle \\
&= \frac{1}{\sqrt{2}}\left(\langle m_1 m_2 | e\hat{r} | m_1^* m_2 \rangle \pm \langle m_1 m_2 | e\hat{r} | m_1 m_2^* \rangle \right) \\
&= \frac{1}{\sqrt{2}}\left(\vec{M}_1 \pm \vec{M}_2 \right).
\end{aligned} \tag{2.48}$$

From the above equations (2.47) and (2.48), it becomes evident that the energy splitting as well as the strength of the respective transition dipole moments are highly dependent on both, the relative orientation of both molecules and the relative orientation of their individual transition dipole moments. As dictated by Kasha's rule, only the low energy state $|l^*\rangle$ is emittive leading to an orientation dependent fluorescence intensity.

The lower left part of figure 2.12 d) schematically shows the dimer: Molecule 1 is fixed in space while the relative orientation of molecule 2 is determined by the tilt angle α and the slip angle relative to the former β. On the right, the respective transition dipole moments are shown. The corresponding energy splitting in units of $(4\pi\epsilon_0)^{-1}$ as a function of β and α is shown in figure 2.12 c). The black lines indicate degenerated states, *i.e.* where $\Delta E = 0$ holds. For a slip stack configuration ($\alpha = 0$) the angle $\beta_\mathrm{m} \approx 35.26°$ is often referred to as the *magic angle*[6] [Par07]. If the orientation of the transition dipoles leads to a weaker fluorescence compared to the monomer emission, the respective area in figure 2.12 c) is noted as "Dark", while areas with higher fluorescence yield are labeled as "Bright". A special case, the slip stack configuration, *i.e.* if $\alpha = 0°$, is considered in figure 2.12 d). The energies with respect to the degenerated states, of $|d_+^*\rangle$ (blue), with $\vec{M}_\mathrm{D}^+ = \sqrt{2}\vec{M}_i$, and $|d_-^*\rangle$ (red), with $\vec{M}_\mathrm{D}^- = 0$ are plotted as a function of the slip angle β. For $\beta < \beta_\mathrm{m}$ the low energy state $|l^*\rangle$ is given by $|d_-^*\rangle$ leading to a dark state in fluorescence as the dipole transition is forbidden. For $\beta > \beta_\mathrm{m}$ however, the sign of the dipole interaction switches, leading to a bright fluorescent low energy state. Crystalline molecular aggregates with a geometry leading to a dark low energy state are called *H-aggregates*, while the counterpart with a bright fluorescence is called *J-aggregate* [KB15]. So far, only electronic transitions have been considered. Nevertheless, an inclusion of vibronic states is possible [Spa10] and will be discussed more thoroughly in the consecutive section 2.3.3.

Electronic excitations for which the excited state resides on a single molecule are often referred to as *Frenkel*-type excitations, in analogy to the exciton theory in crystals discussed in section 2.3.3. However, if the excited electron of molecule 1 is transferred to molecule 2 the resulting state $|m_1^+ m_2^-\rangle$ is called a *charge transfer* (CT) state. If an interaction $\hat{H}_\mathrm{CT}$, similar to that of the Frenkel states, is included in the system's Hamiltonian (2.45), the CT states will lift their degeneracy accordingly, leading to four dimer states, two of Frenkel type and two CT type (compare figure 2.12 b)) [Bar14]. Furthermore, considering additional coupling of the Frenkel and CT states yields overall four adiabatic states $|1\rangle - |4\rangle$ with different ratios of Frenkel and CT "character" [EE17].

As pointed out by Engels and Engel [EE17] treating the molecules as static, even if experimental results are satisfactorily described in some cases, can lead to underestimation of the influence of intra and inter molecular geometrical changes after excitation of the dimer. A simple but representative example is given by the *Excimer*, a neologism composed of the words <u>Exci</u>ted <u>Di</u>mer.

---

[6] In most cases, the magic angle is $54.74° = 90° - \beta_\mathrm{m}$. The phase difference originates from the definition of the slip angle β.

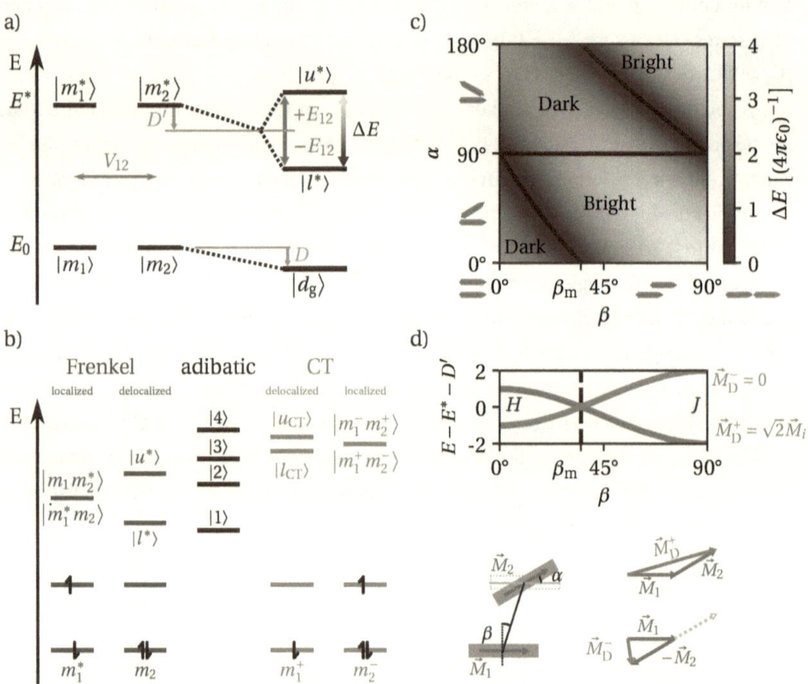

Figure 2.12.: a) Energy levels of monomers and electronic coupling leading to dimer formation. Adapted from [SW07]. b) Extension of the electronic coupling model to charge transfer (CT) states including additional coupling of all states. Excitation schemes for both types are shown below. Adapted from [EE17]. c) Energy splitting as a function of molecular orientation including a coarse divide into dark and bright states referring to the dipole strength of the lower energy component. Black lines indicate $\Delta E = 0$. d) Energy of dimer states for parallel transition dipole moments dependent on the slip angle β showing transition from H- to J-aggregate. Scheme of dipole model as well as transition dipole vector sum for symmetric and anti symmetric states depicted below.

Two molecules characterized by a non attractive ground state potential can form a dimer in the excited state if the excited state resonance energy E_{12} is large enough to rearrange the inter molecular geometry into an attractive configuration [KB15]. In figure 2.13 a) this process is shown along the potential curves of the ground and excited state. Let the inter molecular distance be large enough for the repulsive ground state to not affect the molecules any more ($q = \infty$) like in configuration 1. Photon absorption leads to an excited state with energy E^* on one molecule and the occurring resonance interaction leads to a convergence of the two molecules (configuration 2). The dimer relaxes to the new equilibrium distance q_E of the excited state with energy $E^* - E_{12}$ as shown by configuration 3. After photon emission, both molecules are in the ground state again and are driven apart by the repulsive interaction. For two molecules in a crystal, this process is known as well, however, with the ground state being at the inter molecular equilibrium distance q_0 in the crystal [PS99].

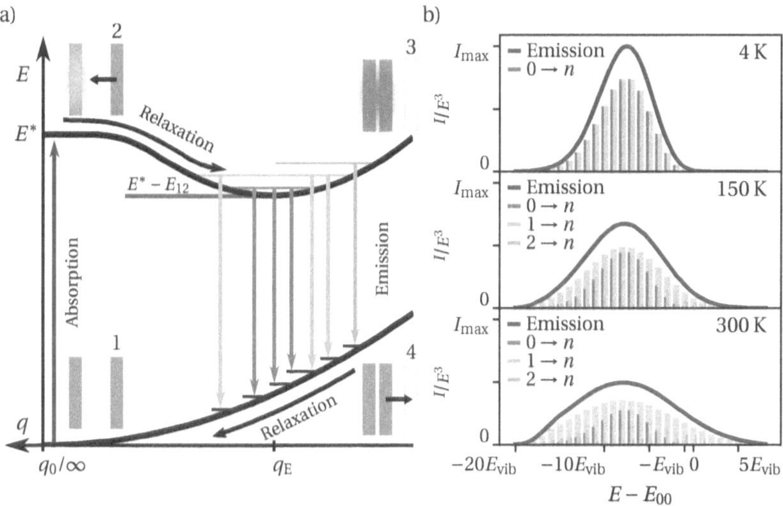

Figure 2.13.: a) Excimer potential and related photo dynamic. The insets 1-4 schematically show the changes in inter molecular geometry. Adapted from [Bir75]. b) Excimer emission spectra for different temperatures relative to the nominal $0-0$ transition. Grey bars indicate the cumulative intensity of the vibronic transitions with the respective energy. Red, cyan and orange bars indicate the intensity part of transitions from the excited state's ground, first, and second vibrational state, respectively.

As reported by Birks [Bir75] the excited state emission can be described by using the Franck-Condon approach of William and Hebb [WH51] for solid state luminescence. The energies of phonon modes in molecular crystals are in the order of $100\,\mathrm{cm}^{-1}$

[SW07] and define the energy magnitude of the inter molecular vibrations of a dimer state, too. The vibrational levels of the excited state are thermally populated. The emission from these levels takes place in a Franck-Condon likne manner, *i.e.* without changing the geometry during the transition between the electronic states. In a crystal, the ground state potential is quantized as well, and the emission spectra is given by

$$W(E;T) = \sum_j \sum_k \underbrace{P(k \cdot E_{D,\mathrm{vib}}, T)}_{\text{Boltzmann probability}} |\langle j \,|\, k \rangle|^2 \, \underbrace{\delta\left(E - \left(E_D^* + E_{D,\mathrm{vib},k} - E_{g,\mathrm{vib},k}\right)\right)}_{\text{defining spectral position of transition}} \qquad (2.49)$$

normalized to the overall electronic transition dipole moment. Here $|j\rangle$ and $|k\rangle$ are the vibrational states with quantum number j and k of the ground and excimer state, respectively. The transition intensity is given by the Franck-Condon factor $|\langle j|k\rangle|^2$, weighted by the respective occupation probability given by a Boltzmann distribution. The Dirac delta function defines the energy of the respective transition, with $E_D^* = E^* - E_{12}$ as the excited dimer energy and vibrational energies $E_{g,\mathrm{vib},j}$ of the ground and $E_{D,\mathrm{vib},k}$ of the excimer state.

In figure 2.13 b) a typical excimer emission spectra is simulated for different temperatures based on a displaced harmonic oscillator model using equation (2.49), with a vibrational energy of 12 meV ($\approx 100\,\mathrm{cm}^{-1}$), and equation (2.39), with a Huang-Ryhs parameter of $S = 8$, to calculate the Franck-Condon factors (for details compare appendix B.1). The abscissa gives the transition energy relative to the 0-0 transition energy E_{00} in units of the oscillator's vibrational energy quanta E_{vib}. The overall all emission spectra composed of gaussian emission components for each transition is displayed in blue. The small energy difference between the vibronic levels leads to a merging of the individual transition peaks known from molecular spectroscopy, yielding a featureless emission characterized only by a broad center peak. The grey bars indicate the transition strength at each energy, while the colored bars show the contribution of the first three vibrational levels of the excimer state to the overall intensity. With rising temperature the thermal occupation of the higher vibronic states leads to a broadening of the peak, while the intensity at the maximum decreases. If the ground state potential becomes increasingly aharmonic for higher energies, the energy difference of the states will become smaller, leading to an emerging high energy tail at high temperatures. This is a typical temperature behavior for excimer emission, shown for example for the dimer emission in pyrene crystals [BKE68].

Certainly, convergence is not the only inter molecular geometry optimization which can lead to excimer formation. Any reaction coordinate leading to a lower energy in the dimer state can lead to a broad featurelesss excimer like emission. For example, after excitation of the perylene bisimide (PBI) dimer, its ground state geometry is distorted as the molecules twist around their inter molecular axis while relaxing into a new excited state geometry with a repulsive ground state potential [Fin+08][EE17].

2.3.3. Crystal Excitons

The preceding section showed that the light matter interaction of a single molecule is drastically altered in proximity of a second molecule. Within a molecular crystal the electronic excitation of an individual molecule will lead to an interaction with its neighboring molecules at adjacent lattice sites. An electronic excitation can be viewed, as Michael Kasha put it, as "a neutral excitation "particle," [*sic!*] consisting of an electron and a positive hole, traveling together through the lattice" [Kas59], or more precise, a quasi-particle called *exciton*. A Coulomb bound electron-hole pair confined to a single lattice site, as in the case of an individual excited molecule in a crystal, can be described by the exciton theory developed by Yakov Frenkel for atomic crystals [Fre31a][Fre31b] and has first been applied to molecular crystals by Alexandr Davydov [Dav48][7]. Such excitons in molecular crystals are usually called *Frenkel excitions* to distinguish them from *Wannier-Mott excitons* found in covalently bound inorganic semiconductors for which the electron is only loosely bound to the positive hole and hence delocalized over several lattice sites [Dem10].

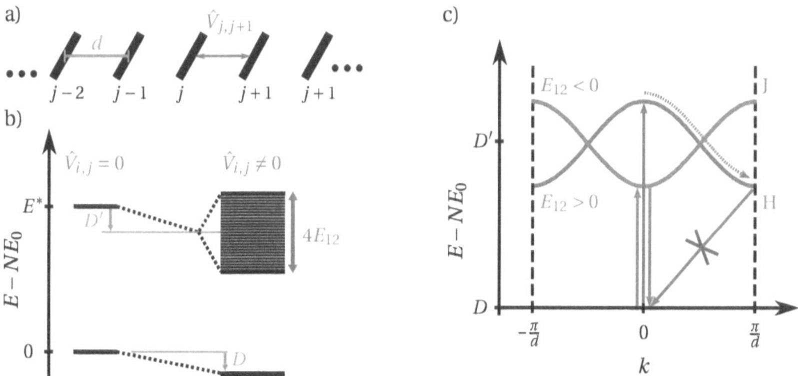

Figure 2.14.: a) Mono molecular 1D linear chain model. b) Energy levels for uncoupled degenerated crystal states and inter molecular coupling. Adapted from [KB15]. c) Exciton band dispersion in the first Brioullin zone for J- and H-aggregates. Solid arrows indicate respective electronic transitions. In an H-aggregate, the direct emission from the lowest energetic state populated by non radiative processes is momentum forbidden. Adapted from [PS99].

---

[7]Historical development from [Kas59]

As a model system, a linear 1D chain of N identical molecules with lattice constant d as shown in figure 2.14 a) shall be considered and is described according to [KB15]. Without loss of generality, let N be an even number. Each molecule is characterized by a ground state $|m_j\rangle$ with energy E_0, an excited state $|m_j^*\rangle$ with energy E^*, and the electronic transition dipole moment $\vec{M}_j = \vec{M}$, while j marks the position on the chain. Choosing N to be large, *i.e.* neglecting the crystal surface or, in case of the 1D chain the 0D endpoints, the Born-von Kármán boundary condition $N+1 \to 1$ can be applied. Similar to the dimer Hamiltonian, the crystal Hamiltonian is given as

$$\hat{H} = \sum_{n=1}^{N} \hat{H}_n + \frac{1}{2} \sum_{\substack{n,m=1 \\ n \neq m}}^{N} \hat{V}_{n,m} \tag{2.50}$$

where $\hat{H}_n$ is the Hamilton operator of the free molecule at position n and $\hat{V}_{n,m}$ is the electronic interaction between the molecules at lattice sites m and n [Dav64]. The overall ground state $|G\rangle$ can be described as a Hartree-product[8] of the individual ground states [EE17]

$$|G\rangle = \prod_{j=1}^{N} |m_j\rangle = |m_1 m_2 \dots m_{N-1} m_N\rangle . \tag{2.51}$$

In the same way the state

$$|E_l\rangle = |m_l^*\rangle \prod_{\substack{j=1 \\ j \neq i}}^{N} |m_j\rangle \tag{2.52}$$

defines an excitation located on the l-th molecule with all other molecules in the ground state [Dav64]. In analogy to the dimer, the excited state of the crystal can be expressed as a linear combination of all possible excitations. Using a Bloch ansatz [SW07] to conserve the crystal symmetry the exciton states are given as

$$|k\rangle = \frac{1}{\sqrt{N}} \sum_{j=1}^{N} \exp\left(ik \cdot dj\right) |E_j\rangle \quad \text{with } k = 0, \pm\frac{2\pi}{Nd}, \pm2\frac{2\pi}{Nd}, \dots, \pm\frac{N}{2}\frac{2\pi}{Nd}, \tag{2.53}$$

where i is the imaginary unit and k is the quasi momentum quantum number of the exciton states $|k\rangle$. [MK64]

---

[8]This approximates the electron-hole pair as a boson. Applying the antisymmetrizer to the Hartree-product considers the fermionic character of the electron and is found in literature as well (*e.g.* [KB15]). As the basic results do not change the boson approximation is used.

Calculating the ground state energy with the crystal Hamiltonian (2.50) yields

$$\langle G|\hat{H}|G\rangle = \left\langle \prod_{p=1}^{N} m_p \left| \sum_{n=1}^{N} \hat{H}_n \right| \prod_{q=1}^{N} m_q \right\rangle + \left\langle \prod_{p=1}^{N} m_p \left| \frac{1}{2} \sum_{\substack{n,m=1\\n\neq m}}^{N} \hat{V}_{n,m} \right| \prod_{q=1}^{N} m_q \right\rangle$$

$$= N E_0 + \underbrace{\left\langle \prod_{p=1}^{N} m_p \left| \frac{1}{2} \sum_{\substack{n,m=1\\n\neq m}}^{N} \hat{V}_{n,m} \right| \prod_{q=1}^{N} m_q \right\rangle}_{=D}. \tag{2.54}$$

The sum of the ground state energy of the unbound molecules is shifted by the molecules' mutal van der Waals interaction, resulting in the total energy shift D. As the van der Waals interaction between the neutral constituents of a molecular crystal usually leads to bonding, D is defined as a negative contribution, as shown in figure 2.14 and known as the *gas-to-crystal shift*.

Applying the crystal Hamiltonian $\hat{H}$ to an excitonic state $|k\rangle$ two interaction contributions can be separated. One is the van der Waals interaction of the excited state with its adjacent molecular ground states

$$D' = \frac{1}{N} \sum_{p=1}^{N} \left\langle E_p \left| \frac{1}{2} \sum_{\substack{n,m=1\\n\neq m}}^{N} \hat{V}_{n,m} \right| E_p \right\rangle \tag{2.55}$$

and is usually a negative, *i.e.* attractive, contribution. The second contribution is the resonance interaction of two excited states and can be reasonably approximated by taking only next neighbor interactions $\hat{V}_{j,j\pm 1}$ into account. Thus, it is given by

$$\frac{1}{N} \sum_{q=1}^{N} \left(\underbrace{\exp(-ik\cdot d)\langle E_q|\hat{V}_{q,q+1}|E_{q+1}\rangle}_{=E_{12}} + \underbrace{\exp(ik\cdot d)\langle E_q|\hat{V}_{q,q-1}|E_{q-1}\rangle}_{=E_{12}} \right)$$

$$= 2E_{12}\left(\exp(-ik\cdot d) + \exp(ik\cdot d)\right)$$

$$= 2E_{12}\cos(k\cdot d). \tag{2.56}$$

The resulting interaction energy shows that the N-fold degeneration of the excitons is lifted and with $N \to \infty$ an exciton band of width $\Delta E = 4E_{12}$ is formed where E_{12} is the resonance interaction energy between two neighboring molecules. Together with the remaining molecular energy operator $\sum_{j=1}^{N} \hat{H}_j$ the overall energy of the $|k\rangle$-exciton reads

$$E(k) = (N-1)E_0 + E^* + D' + 2E_{12}\cos(k\cdot d). \tag{2.57}$$

The changes in the energy level diagram from the non interacting linear chain to an interacting one is shown in figure 2.14 b).

Electronic transitions $|G\rangle \rightarrow |k\rangle$ can be described according to [SB16] in the following way: The transition dipole moment operator is the sum over all molecular transition dipole operators. The squared transition dipole moment determining photon absorption and emission rates is given by

$$\left|\left\langle G \left| \sum_{j=1}^{N} \hat{\mu}_j \right| k \right\rangle\right|^2 = N|\vec{M}|^2 \delta(k) \tag{2.58}$$

including the momentum conserving selection rule restricting transitions strictly to the Γ-point. As discussed for the dimer in the point dipole approximation, the nearest neighbor coupling E_{12} can lead to an upward or downward shift in energy for the symmetric state corresponding to the $|k=0\rangle$ exciton state. There are various mechanisms dominating on different length scales which can become relevant for the energetics in a crystal including, amongst others, a resonance energy contribution by orbital overlap or dipole-dipole interaction. However, for all couplings the cases $E_{12} < 0$ and $E_{12} > 0$ can be distinguished. Figure 2.14 c) shows the dispersion relations in the first Brioullin zone for $E_{12} < 0$ (red) and $E_{12} > 0$ (blue). For $E_{12} < 0$ photo excitation leads to the lowest energetic state from which emission with an N-times faster emission rate (ignoring frequency shifts with respect to monomer emission) compared to the monomer is possible. Such a molecular configuration is called J-aggregate as before. If the resonance interaction is positive the $|k=0\rangle$ state resides at the top of the exciton band. According to Kasha's rule, a photo excited state relaxes to the band minimum at the Brioullin zone boundary $k = \pi/d$ from where, according to (2.58), photon emission is forbidden leading to a dark state, and the corresponding molecular configuration is again called H-aggregate.

The model of the linear chain can be expanded to two translationally nonequivalent molecules a and b in a unit cell, like in figure 2.15 a). The presented theoretical description is based on [Dav48], [Dav64], and [SW07]. An intra unit cell coupling $\hat{V}_{ab}$ between the molecules a and b and an inter unit cell couplings, $\hat{V}^a_{m,n}$ and $\hat{V}^b_{m,n}$, along the chain for each molecule have to be taken into account. The respective resonance energies are termed E_{ab}, E^a_{12}, and E^b_{12}. While the inter unit cell coupling leads to a band dispersion for each type of molecule, the intra unit cell interaction splits the bands as shown in figure 2.15 b). This splitting of the exciton bands is called *Davydov splitting* and the two components are referred to as *Davydov components* or *Davydov states*. For unit cells containing two identical molecules, *e.g.* anthracene [SW07], the two Davydov states can be formed by a linear combination of the individual excitonic states

$$|k; \pm\rangle = \frac{1}{\sqrt{2}} \left(|k; a\rangle \pm |k; b\rangle \right). \tag{2.59}$$

The electronic transitions can be treated analogously to the molecular dimer from section 2.3.2 leading to two perpendicular transition dipole moments $\vec{M}_1$ and $\vec{M}_2$ obeying the $k = 0$ selection rule for each band.

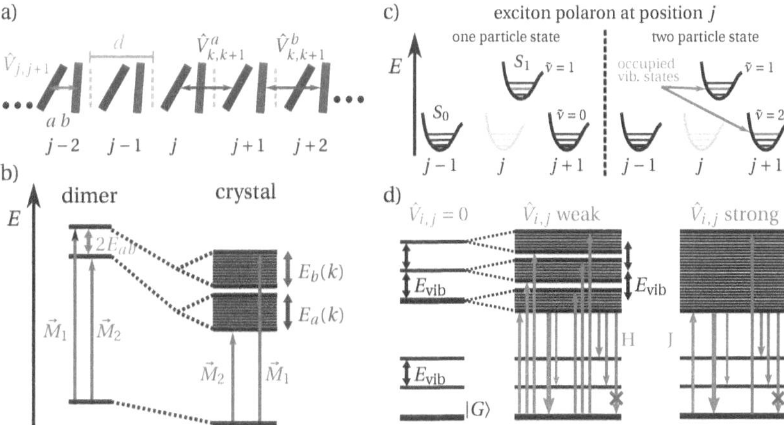

Figure 2.15.: a) Linear 1D chain model with two translationally nonequivalent molecules in the unit cell. b) Energy level diagrams for the unit cell dimer and for the corresponding crystal exctions including the optical transitions. Adapted from [SW07]. c) Schematic representation of the exciton polaron model on a linear 1D chain. Molecular potentials are represented in a displaced harmonic oscillator model with vibrational levels $\tilde{\nu}$. The molecule at position j in the S_1 state represents the exciton. In a one particle state the adjacent molecules are in their vibrational ($\tilde{\nu} = 0$) and electronic ground state S_0, while for a two a particle state one additional adjacent molecule is vibrationally excited ($\tilde{\nu} > 0$). d) Optical transitions and energy levels for the two coupling limits. The superradiant 0–0 transition for a J-aggregate is shown as a thick red arrow. c) and d) adapted from [Spa10].

So far, the molecules have been treated as a rigid two level system. However, as discussed in section 2.3.1 a molecular excitation leads to an elongation of the equilibrium distance and is usually accompanied by vibrational excitations. In molecular crystals the adjacent molecules will be vibrationally excited as well. As the excitation diffuses through the crystal the vibrational energy and hence, the deformation at the exciton's position and its adjacent lattice sites behaves accordingly. This is called an *excitonic polaron* [Spa10]. A short summary of the basic results will follow, for a detailed discussion it is referred to [SB16] and [Spa10].

The exciton polaron model system described by Frank C. Spano [Spa10][SB16] constitutes of a linear chain of molecules with one (dominant) vibrational mode energy E_{vib}. The photon absorption and emission is described by two types excited states which are schematically shown in figure 2.15 c). The *one particle state* comprises a vibronic excitation on a single molecule while its adjacent molecules are in their respective molecular ground state S_0. For a two particle state, in addition to the vibrationally excited exciton, an adjacent molecule is vibrationally excited as well. Higher order multi particle states have negligible effect on vibronic transitions and hence, these two types of exciton polarons are sufficient to describe photon absorption and emission. The energetic structure can be described by the *Holstein Hamiltonian*, which is composed of the exciton hamiltonian of the linear chain (2.50) and an on-site vibronic coupling. In figure 2.15 d) on the left, the energy levels including the vibrational levels of the uncoupled molecules in the crystal are shown. When resonance coupling is considered, the individual vibrational levels form exciton bands, separated by the vibrational energy E_{vib} as shown in the center of figure 2.15 d). The gas-to-crystal shift has been neglected for simplicity. In the strong coupling limit, *i.e.* if the resonance energy E_{12} is much larger than the molecular reorganization energy SE_{vib}, the exciton resonance exchange with its neighboring molecular states is so fast that, in principle, no vibrational relaxation takes place and the excitons and phonons can be treated independently as described by the Born-Oppenheimer approximation leading to the formation of a single exciton band. As a result, the optical absorption, as indicated in figure 2.15 d), is dominated by the $k = 0$ momentum selection rule. For an H-aggregate the optical accessible state sits at the top of the band, while for J-type aggregation preferentially the lowest energy state is excited. This holds for vibronic bands of small coupling as well as for exciton bands of strong coupling. For small resonance coupling ($E_{12} \leq SE_{\text{vib}}$), the absorption spectra show a vibronic spacing, however, depending on the coupling strength, the intensity distribution of the vibronic transitions differs from the Poisson distribution (2.40) for monomers. For strong resonance coupling the absorption is reduced to a single line, corresponding to the linear chain exciton as discussed before. According to Kasha's rule, the emission starts from the lowest excited state in the lowest (vibronic) band as shown in figure 2.15 d). Contrary to H-aggregates of rigid molecules, photon emission is possible by transitions into a vibrationally excited state of the electronic ground state, as the creation of a vibron ensures conservation of momentum. However, a direct transition to the vibrational ground state is still forbidden. For a J-aggregate the direct emission to the ground state, in analogy to the single molecule termed 0–0 transition, is allowed and its transition strength scales with the number of molecules N as indicated by the transition dipole moment (2.58). Lower lying, consecutive vibronic transitions do not benefit from the exciton delocalization [Spa10]. A general relation for the reduced intensity (*i.e.* $I(\omega_{em})/\omega_{em}^3$, *c.f.* section 3.4.1) relating the 0–0 transition and the 0–1 transition can be derived [SY11]:

$$\frac{I_{0-0}}{I_{0-1}} = \frac{N_{coh}}{S}.$$

(2.60)

Here, S is the Huang-Rhys parameter defining the reorganization along the vibrational mode. In the general case, the number of monomers N is replaced by the *number of coherently excited molecules* N_{coh}, *i.e.* the number of molecules over which the exciton is actually delocalized and hence, the amount of states from which the crystal exciton state defined in equation (2.28) is formed. This modifications considers exciton confinement on only several monomers rather than delocalization over the whole chain due to thermally induced decoherence (dynamic disorder) and chain defects (static disorder) [SY11]. The increase in the 0–0 transition intensity with the number of coherently excited molecules N_{coh} is referred to as *superradiance*[9] [Spa10].

2.3.4. Charge Transfer Complexes

As already mentioned in the discussion of a dimer comprising two identical molecules (*c.f.* section 2.3.2), besides Frenkel excitons there are also charge transfer states, for which the excited electron is transferred to the LUMO of an adjacent molecule. Although of importance in aggregates of identical molecules [EE17], these states dominate the photophysics of mixed aggregates comprising different molecules with HOMO and LUMO offset as shown in figure 2.16. The molecule with the energetically higher lying HOMO and LUMO is defined as the *donor* and the molecule with the respective lower lying levels, is called *acceptor*. Such a configuration can be approximated by Mulliken theory [Mul52], where the interaction between the ground and excited states of the donor and the acceptor defines the resulting states of the donor-acceptor (D-A) configuration and an electronic transition results in a (partial) charge transfer between both entities [Bel+17]. The excited states of such a D-A configuration are often referred to as *charge transfer (CT) states* and, *inter alia*, play an important role in organic optoelectronic devices [DSD10][Van16] making them a subject of intensive research [Luk20][Bro+18].

---

[9]This phenomena has to be distinguished from the *superradiance* introduced by Robert H. Dicke [Dic54], covering coherent radiation from spontaneous emission in a molecular gas for which the transition intensity scales with N^2.

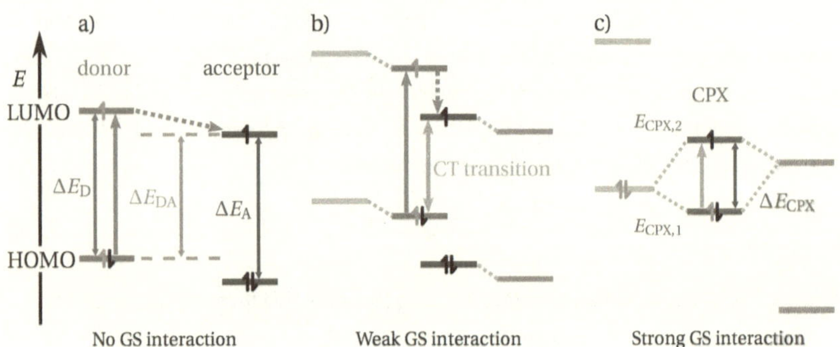

Figure 2.16.: Energy levels for a D/A complex distinguished by the ground state (GS) interaction strength. a) No GS interaction with D/A band gap similar to donor and acceptor band gap. After excitation of the donor, the electron is transferred to the acceptor's LUMO. b) For weak GS interaction and a small D/A bandgap compared to ΔE_D or ΔE_A a partial charge transfer is present in the ground state and the CT state can be excited either from the excited donor or directly via photon absorption. c) For a strong ground state interaction, *i.e.* when $\Delta E_{DA} \approx 0$, a charge transfer complex (CPX) is formed and the donor's HOMO and acceptor's LUMO hybridize which allows excitation via photon absorption. Adapted from [Bel+17].

Belova *et al.* [Bel+17] classify the D-A systems by the strength of their ground state interaction and distinguish the three possible cases presented in figure 2.16 with increasing interaction strength from left to right. According to this classification, the comparison of the formed D-A energy gap ΔE_{DA}, *i.e.* the energy difference between the donor's HOMO and the acceptor's LUMO, with the molecular HOMO-LUMO energy gaps of the donor ΔE_D and the acceptor ΔE_A, can be used as an indicator for the interaction strength of ground states.

If the energetic offset is marginal, the D-A energy gap is close to one of the molecular energy gaps ($\Delta E_{DA} \approx \Delta E_D$ or ΔE_A). In this configuration a ground state charge transfer is unlikely [Bel+17]. Upon optical excitation of the donor an integer charge transfer of the excited electron from the donor to the acceptor's LUMO can be a favorable relaxation process [Van16] even if the difference between the donor's LUMO and the acceptor's LUMO is small [Kaw+15]. This leaves a spatially separated but Coulomb bound electron hole pair with the charge transfer character. The additional binding energy due to Coulomb interaction between the electron and hole, screened by the surrounding dielectric medium, is usually estimated to be around 0.5 eV to 1.0 eV for a generic organic semiconductor [KB15][Van16]. The excited donor state relaxes to the CT state within 10 fs to 100 fs [Van16] making it an ultra fast photophysical relaxation process (compare table 2.1). The electron transfer is thought to be a coherent process, during which the charge oscillates between the donor and the acceptor with a $\approx 25\,$fs period before fully relaxing [Fal+14][JRB17]. Calculations suggest that in crystalline D-A heterojunctions, charge delocalization over several molecules leads to an overall lowering of the CT state energy while reducing the exciton binding energy simultaneously [Yan+14]. A hole transfer from an excited acceptor to the donor molecule is possible as well and has been observed even for a vanishingly small HOMO-HOMO offset [Li+19].

In the case of weakly coupled molecules, *i.e.* with $\Delta E_{DA} < \Delta E_D$ or $\Delta E_{DA} < \Delta E_A$, a partial D-A charge transfer is possible in the complex's ground state. Although the molecular orbitals do not significantly vary with respect to the monomers, the HOMO and LUMO levels can be shifted compared to the molecular levels [Sch+16][Bel+17]. This is shown in figure 2.16 b) together with the possible electronic transitions (solid lines) and relaxations (dotted lines) leading to the charge transfer state. Due to the ground state interaction, a direct excitation and radiative relaxation of the charge transfer state is possible, even though the oscillator strength is usually low compared to molecular processes. [Bel+17]

If the donor's HOMO level and the acceptor's LUMO level are energetically close, *i.e.* for $\Delta E_{DA} \approx 0$, a strong inter molecular coupling is observed [Bel+17], leading to hybridization of the molecular orbitals [Sal+12] as depicted in figure 2.16 c). The lower lying hybrid orbital of the emerging *charge transfer complex* (CPX) $E_{CPX,1}$ is occupied by the donor's HOMO electrons [Sal+12] and can be a doping mechanism to increase the charge carrier density in organic semiconductors if the complex's energy gap is small enough to allow ionization at room temperature [Mén+13][Mén+15]. The resulting band gap can be described by means of a Hückel approach [Mén+13][Mén+15]

$$\Delta E_{CPX} = E_{CPX,2} - E_{CPX,1} = \sqrt{\left(E_{D,H} - E_{A,L}\right)^2 + 4\beta} \tag{2.61}$$

with β being the resonance integral (compare section 2.1.2 on page 13). Optical transitions exciting an electron from the CPX HOMO $E_{CPX,1}$ to the CPX LUMO $E_{CPX,2}$ have been observed in absorption studies [Mén+13][Duv+18]. Even though luminescence is a plausible relaxation process, it has not to been reported in literature for strongly interacting complexes so far.

2.4. Organic Light Emitting Diodes

Organic Light-Emitting Diodes are one of the most successful implementations of organic semiconductors in consumer oriented electronics [You19]. The underlying physical effect, *electroluminescence, i.e.* photon emission due to the recombination of charge carrier of opposite polarity, was first demonstrated for organic molecular semi-conductors using anthracene single crystal [PKM63][HS65] but the high driving volt-ages of 100 V to 1000 V in combination with a low photon yield ruled out potential application at this time. From this first demonstration, it took more than 20 years until Ching Wan Tang and Steven van Slyke demonstrated a device with sufficient light output at reasonable opperating voltages under 10 V based on the molecule 8-hydroxyquinoline aluminium (Alq_3) [TV87]. These OLEDs used fluorescence, *i.e.* the radiative decay of singlet excitons, for light generation which is, if induced from charge carrier recombination, usually limited to an internal quantum efficiency of 25 % as, according to spin statistics 75 % of the generated electron-hole pairs are triplet exci-tons [BBM01]. As mentioned in section 2.3.1 triplet states in organic semiconductors usually have low phosphorescence rates, but Marc Baldo *et al.* demonstrated that an internal quantum efficiency of 100 % in OLEDs is possible using spin orbit cou-pling [Bal+98]. Even though the generally low triplet energy makes development of blue phosphorescent emitters challenging it is possible to achieve high yields in all re-gions of the visible spectrum, making the development of efficient white light emitting diodes possible [Rei+09]. In 2012 the concept of *thermally activated delayed fluores-cence* was introduced [Uoy+12], which uses highly efficient reverse inter system cross-ing to recover the generated triplet states for fluorescence, and hence, constitutes a different pathway to almost 100 % internal quantum efficiency.

This section will give a short introduction to the field of OLEDS by describing the underlying physics of metal-organic contact interfaces for charge injection, giving an overview on the description of electrical currents in organic semiconductor diodes, and, finally, explaining the working principle of such devices by means of a simple one layer model. The section is mainly based on [BBM01], [SW07], [KB15], and [HDT16] which are referred to for a more detailed discussion.

The transport levels of holes and electrons will be treated as if corresponding to the respective molecular HOMO and LUMO levels in the solid. This is a common simplifi-cation in literature [BBM01][Ish+99]. However it has to be kept in mind that the actual transport levels are shifted from the neutral HOMO and LUMO levels as the introduc-tion of a hole or an electron produces a molecular radical cation or anion, respectively. This instantaneously leads to a rearrangement of the electrons and thus to a polariza-tion of the surroundings, as well as, on longer timescales of, to a rearrangement of the molecular geometry, which in return leads to a shift of the molecular energy [KB15].

2.4.1. Contacts and Interfaces

Figure 2.17 a) depicts the energy diagram of an organic semiconductor positioned between two metals aligned at the common vacuum level. Based on the representation of transport levels in inorganic semiconductors all levels are shown as a continuous level throughout the organic layer, although, in reality, they are rather a distribution of individual molecular levels due to the inherent energetic disorder of the polycrystalline or amorphous structure [KB15]. The semiconductor's ionization energy is denoted by I_O, while A_O represents its electron affinity. The work functions Φ and Fermi energies E_F of the metals are labeled Φ_{M1}, Φ_{M2}, $E_{F,M1}$, and $E_{F,M2}$, respectively. If the metals are connected, charge carriers will flow from the low work function metal M1 to the high work function one to equate the Fermi level offset (*cf.* figure 2.17 b)). The now charged metal contacts generate an electric field within the organic semiconductor which acts as a dielectric due to the lack of internal free charge carriers. This *built in potential*

$$V_{\mathrm{BI}} = \frac{1}{e} \left(\Phi_{M2} - \Phi_{M1} \right) \tag{2.62}$$

is determined by the work functions of the metal contacts and needs to be compensated before charge carrier injection into, and subsequent transport through, the organic semiconductor is possible.

Figures 2.17 a) and b) show an injection barrier for holes or electrons from the metal into the HOMO or LUMO level of the organic semiconductor. In the Schottky-Mott model, the vacuum level of the metal and the semiconductor align at the interface and the injection barrier is calculated as the difference between the work function of the metal and the ionization potential for hole or the electron affinity for electrons, respectively [SW07]. However, this model is over simplified in most cases as dipole formation at the interface can lead to quite significant shifts of the vacuum level in either direction which, in some cases, exceed even 1 eV [Ish+99]. For some interfaces, due to charge accumulation, a Fermi level pinning accompanied by band bending across a few molecular layers has been observed [Bre+15]. Strong dipole interactions leading to vacuum level offsets are not only found at metal/organic interfaces but can be present also at organic/organic interfaces most likely induced by strong charge transfer complexes [Bre+15][Váz+05]. This is only a rather brief treatment of the various interface phenomena, as a full treatment can be found elsewhere [Ish+99][SW07][KB15].

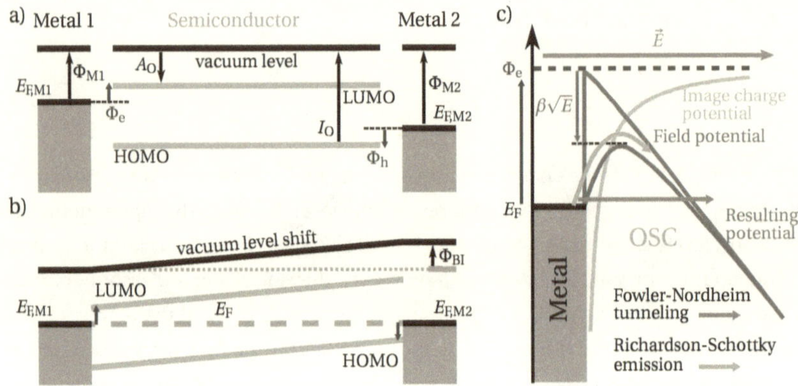

Figure 2.17.: Energy level diagram of metal/organic semiconductor (OSC)/metal interface stack representing a simple one layer OLED device. a) Before contact all levels align at the common vacuum level. b) Upon contact, the Fermi level alignment of the metal layer causes a built in potential Φ_{BI}. For electron and hole injection the respective band offsets Φ_e and Φ_h have to be overcome. c) The electron injection barrier Φ_e at a metal/OSC interface in an applied external electric field $\vec{E}$. The two main processes for charge carrier injection, the thermionic Richardson-Schottky emission (green) and the Fowler-Nordheim tunneling (red), are indicated. Adapted from [SW07].

2.4.2. Charge Carrier Injection and Transport

A desirable situation for charge injection into the organic semiconductor would be an *Ohmic contact*, where the amount of free charge carriers in the contact and the interface is high enough to be regarded as an infinite reservoir. However, this is usually not the case, as most interfaces form an injection barrier for charge carriers, limiting the injection and thus, the amount of available free charge carriers at the interface. In figure 2.17 c), the electron injection barrier at a metal/organic semiconduct is shown in an applied external electric field. The discussion for electron extraction, *i.e.* hole injection, can be treated equally.

Charge carrier injection is governed by two main processes, a thermionic emission of charge carriers across an injection barrier or a tunnel process through this injection barrier [BBM01]. The former process is described by the *Richardson-Schottky equation*

$$j_{RS} = \underbrace{\frac{4\pi e m^* k_B^2}{h^3}}_{A^*} T^2 \exp\left(-\frac{\Phi_e - \beta\sqrt{E}}{k_B T}\right), \tag{2.63}$$

where A^* is the *Richardson constant* and m^* being the effective mass of the charge carriers. In argument of the exponential term, $\beta\sqrt{E}$ describes the reduction of the injection barrier Φ_e by superposition of the potential of the applied electric field (*cf.* figure 2.17 c) purple) with that of an residual image charge in the metal contact caused by an electron already injected into the organic semiconductor (*cf.* figure 2.17 c) grey) resulting in the *reduced injection barrier* $\Phi_B = \Phi_e - \beta\sqrt{E}$ (*cf.* figure 2.17 c) green) [BBM01] [SW07]. In figure 2.17 c) the red arrow indicates a tunnel process at the Fermi energy of the metal through the injection barrier modified Φ_B the applied field $\vec{E}$. Such a tunnel process is described by *Fowler-Nordheim tunneling* and yields the following electrical current density [FN28] [SW07]

$$j_{FN} = \frac{A^*}{\Phi_e}\left(\frac{eE}{\alpha k_B}\right)^2 \exp\left(-\frac{2\alpha\Phi_e^{3/2}}{3eE}\right) \tag{2.64}$$

with $\alpha = 4\pi\sqrt{2m^*}/h$. If the charge carrier injection is the limiting process over in the applied field range, the current density in the device is described by one of these processes.

For Ohmic contacts or, if at least one contact injects significantly more charge carriers into the device than intrinsically present at thermal equilibrium, the contact region quickly becomes flooded with charge carriers. Due to the low mobility in the organic semiconductors the charge accumulates and builds up a space charge region with the resulting field determining the charge transport. This results in a *space-charge limited current*. In an idealized organic semiconductor there is just a single transport level on which the charge carriers move with their intrinsic mobility μ_0. Upon applying an electric field, the current density through the device of thickness d is given by the *Mott-Gurney equation*

$$j = \frac{9}{8}\epsilon\epsilon_0\mu_0\frac{V^2}{d^3}. \tag{2.65}$$

where V is the applied voltage, and ϵ and ϵ_0 are the absolute and the vacuum permittivity, respectively. However, this assumption can only serve as a model, as the reality turns out to be more complex. If the spatial continuity of this transport level is disrupted by trap states which immobilize the quasi-free charge carriers by capturing them in a potential well with an energetic barrier E_T, with respect to the transport level, the overall current is decreased. The impact on the current density is mostly determined by the ratio of the trap density N_T and the density of states N_c at the transport level. In this case, the current density can be described by the Mott-Gurney equation (2.65) replacing the intrinsic mobility μ_0 with an *effective mobility*

$$\mu_{\text{eff}} = \mu_0 \left(1 + \frac{N_\text{t}}{N_\text{c}} \exp\left(\frac{E_\text{T}}{k_\text{B}T}\right)\right).$$

(2.66)

If the current density surpasses a certain threshold above which all trapped states are occupied, *i.e.* leading to an effective trap density $N_\text{T} = 0$, the current density suddenly increases as the mobility returns to the intrinsic value. The device operates in the so called *trap filled limit* of the fully space-charged limited current regime.

The above cases are, of course, only valid if a constant intrinsic mobility μ_0 exists, if homoenergetic charge carrier traps interrupt an otherwise constant transport level, and if charge carrier diffusion can be neglected. A more thorough treatment can be found in [BBM01] and [KB15]. Moreover, is has been recently shown that for electrons, trapping in potential wells might not be the limiting process but rather reflection at grain boundaries leads to a hindered charge carrier transport [Vla+18]. Albeit, the microscopic charge carrier transport mechanisms have not been fully disclosed yet, as there are various competing processes to be considered, a more exhaustive discussion can be found, for example, in recent reviews such as [LCB16] and [Sch+20].

2.4.3. OLED Working Principle and Charge Carrier Recombination

The working mechanism of an OLED can be split into four essential steps as shown in figure 2.18 [BBM01]. For efficient ambipolar charge carrier injection, contacts for electron and hole injection differ in their work functions leading to a built in potential. For charge carrier injection and transport the applied voltage needs to compensate this built in voltage and restore the flat band condition. After the applied voltage is large enough for charge carrier transport, the charge carriers still have to overcome their respective injection barrier to reach their transport level in the organic semiconductor (step 1). Afterwards, electrons and holes are transported through the device (step 2). The measured device characteristics dependent on the major limiting process, *i.e.* injection or space-charge limitied current, and can vary for different voltage strengths and device configurations[KB15]. The essential precursor for electroluminescence is the formation of an exciton by an unbound hole and an unbound electron (step 3). The exciton can either be located on a single molecule or can have multi-molecule character, as in case of a CT exciton. The radiative relaxation to the ground state releases a photon (step 4). [BBM01]

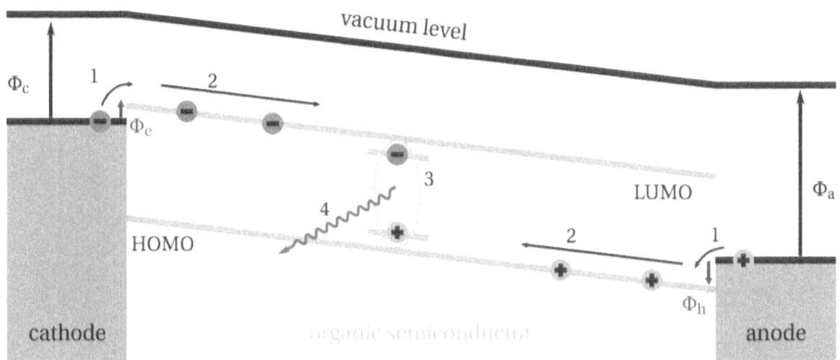

Figure 2.18.: OLED working principle: (1) Charge carrier injection into the semiconductor by overcoming the injection barriers Φ_e and Φ_h. (2) Charge carrier transport across the organic semiconductor layer towards the opposed electrode. (3) Exciton formation by coulombic capturing of a close-by electron and hole. (4) Radiative decay of the exciton leads to photon emission. Adapted from [BBM01].

The exciton formation can be regarded as a bimolecular process and is successfully described by Langevin-theory [AB95a][AB95b]. The critical capture radius is reached when the Coulomb interaction between the two oppositely charged charge carriers surpasses their thermal energy and is given by

$$r_c = \frac{e^2}{4\pi\epsilon\epsilon_0 k_B T}$$

(2.67)

wheras the exciton formation rate is reads

$$\gamma_{eh} = \frac{e\left(\mu_h + \mu_e\right)}{\epsilon\epsilon_0}$$

(2.68)

with μ_h and μ_e are the respective hole and electron mobilities.[PS99]

3. Materials and Methods

3.1. Materials

This section gives a short and by no means exhaustive overview on the used materials. The two main material systems, the donor-acceptor system pentacene-perfluoropentacene, and zinc(II)phthalocyanine are briefly introduced and put into an introductory scientific context. Subsequently, electrode and transport materials used for prototypical device studies are described relating their most important properties.

3.1.1. Pentacene and Perfluoropentacene

Pentacene (PEN) is a polycyclic aromatic hydrocarbon comprising five annulated benzene rings (*c.f.* 3.1 a)). It has first been synthesized in 1930 by Clar *et al.* [CJ30]. It is widely studied as a hole conducting material with mobilities in single crystals ranging from $1\,\mathrm{cm^2\,V^{-1}\,s^{-1}}$ to $5\,\mathrm{cm^2\,V^{-1}\,s^{-1}}$ [LRP06][TOM12] with reported values of up $40\,\mathrm{cm^2\,V^{-1}\,s^{-1}}$ [Jur+07] at room temperature. Various solid state phases are known for pentacene and the different polymorphs can be distinguished by their lattice spacing along the [001] direction [Mat+01]. The molecular arrangement in the crystallographic (ab)-plane presents as a herringbone pattern for all polymorphs [Mat+03a]. Pentacene single crystals usually adopt the $14.1\,\text{Å}$ [Mat+01][Sie+07] (001) spacing with an enantiotropic phase transition at 463 K to a high temperature phase, also known as the *Campbell*-phase [CRT61], with a $14.5\,\text{Å}$ spacing along [001] [Sie+07]. Both unit cells are triclinic with the space groups $P\bar{1}$ and $P1$, respectively. Depending on the substrate and the preparation conditions, a metastable thin film phase with a d_{001}-spacing of $15.4\,\text{Å}$ can be prepared [Mat+03b] which can be transformed to the high temperature phase via thermal treatment [Mat+03b] or solvent exposure [Gun+99]. It has been demonstrated that the solid state ionization energy I_{c} [Duh+08][Yos+15] as well as the electron affinity A_{c} [Yos+15] which are associated with the HOMO and LUMO levels in the crystal depend on the molecular orientation. For pentacene in a standing up/lying down orientation the HOMO level is situated at $-4.90\,\mathrm{eV}/{-5.45}\,\mathrm{eV}$ and the LUMO energy is found at −2.35 eV/ −3.14 eV, respectively [Yos+15], as shown in the energy diagram in figure 3.1 b).

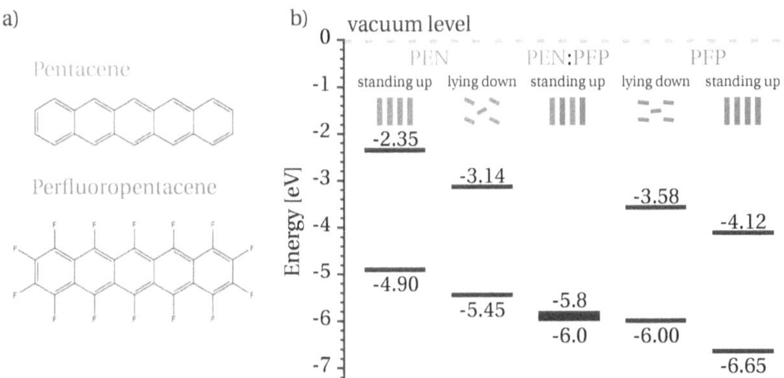

Figure 3.1.: a) Molecular structure of pentacene (PEN) and perfluoropentacene (PFP). b) Ionisation energies and electron affinities as HOMO and LUMO levels for standing up/lying down orientation of PEN and PFP [Yos+15] as well as the range of the HOMO level of a mixed PEN:PFP film of upright standing molecules [Sal+08b].

Replacing the pentacene's hydrogen ligands by fluorine yields the preferential electron conductor *perfluoropentacene* (PFP), first synthesized and characterized by Sakamoto *et al.* [Sak+04]. The crystal structure is similar to the one of pentacene, with an in plane herringbone pattern, although the lattice spacing along the crystallographic [100] direction $d_{100} = 15.5$ Å, resembling the long molecular axis, is larger compared to pentacene and the monoclinic unit cell is of higher symmetry (space group P 2_1/c (14))[Sak+04]. Furthermore a thin film phase with slightly larger lattice spacing of (15.75 ± 0.05) Å has been reported on SiO_2 [Sal+08a]. The crystal HOMO and LUMO energies, as shown in the energy diagram in figure 3.1 b), for the standing up/lying down orientation are −6.65 eV/−4.12 eV and −6.00 eV/−3.58 eV, respectively [Yos+15]. As known for various other molecules, fluorination leads to an overall downward shift of the HOMO and LUMO levels without altering the band gap[1] too much [Med+07]. The band gap $E_g = |A_c - I_c|$ for pentacene $E_{g,PEN}$ ranges from 2.3 eV to 2.6 eV, depending on molecular orientation, and is only slightly altered upon fluorination and hence, $E_{g,PFP}$ lies within the range of 2.4 eV to 2.5 eV. However, an overall shift to lower energies for both, the HOMO and the LUMO energy, is obvious (*c.f.* figure 3.1 b)) [Yos+15].

---

[1] Even though often applied, one has to distinguish the optical band gap, *i.e.* absorption and emission energy, from the band gap defined by the ionization energy and the electron affinity, as the former is determined in a neutral molecule while the latter refers to a radical cation or anion in a dielectric environment.

This makes pentacene-perfluoropentacene an interesting donor-acceptor system which has been largely studied. Mixed films with a 1:1 molar ratio crystallize as a co-crystal [Hin+11] in a $P\bar{1}$ symmetry [Kim+19]. In the co-crystal, the HOMO level lies between those of the individual molecules in a range from 5.8 eV to 6.0 eV [Sal+08b], as shown in figure 3.1 b). Furthermore, the direct photon absorption by a charge transfer state at 1.6 eV [Bro+11] and its emission at 1.4 eV [Ang+12], being well in the spectroscopically detectable energetic region, make these molecules a model system for a charge transfer complex and, thus, are of a fundamental academic as well as technological interest.

3.1.2. Zinc Phthalocyanine

Phthalocyanines (Pc) are macrocyclic molecules based on a porphyrazine ring extended by benzene rings on the pyrrole units [MT63]. Its first but unintentional synthesis as a byproduct has been achieved by Braun and Tcherniac in 1907 [BT07]. Since their discovery phthalocyanines have found various fields of applications [Gre00]. They are used on large scales as colorants for printing inks, varnish or paints with a production value of over one billion U.S. dollars [Wöh+12][Gre00], are discussed as photosensitisers for photodynamic cancer therapy [Bon95][Lo+20] and have been thoroughly studied as a donor material in organic photovoltaic cells [Bre+15][Bru+10], to name just some areas of application. "The phthalocyanines" comprise a group of molecules based on the phthalocyanine molecule. The subgroup of the metal phthalocyanines (M-Pc) are derivatives in which the phthalocanine's two inner hydrogene atoms are replaced by a metal atom such as zinc (*cf.* figure 3.2) without heavily altering the structural properties in the solid state although the electronic properties can be heavily influenced [Dav82] [LS01].

It has been largely accepted that there are at least three polymorphic forms in most phthalocyanine solid states [MT63][Wöh+12] with one thermodynamically stable β phase in single crystals [Bro68] and various metastable phases [Ass65][KUS68], most commonly the α phase, preferentially stabilizing in crystalline thin films [Ber+00][Heu+00]. An $\alpha \rightarrow \beta$ phase transition can be induced thermally above 500 K [SM68][HG92] or by solvent exposure[KUS68].

Zinc(II)phthalocyanine (ZnPc) is a M-Pc with zinc as central metal atom. It has superior optoelectronic properties compared to other M-Pc's as its lowest excited state is not quenched by ligand metal charge transfer states due to zinc's fully occupied 3d orbitals [Bru+10]. Polymorphism, as for most Pc's, is also found in ZnPc [SHL07] and the $\alpha \rightarrow \beta$ transition can be induced thermally [Gaf+10] as well as via solvent exposure [IKU80]. It has been established that the crystal structure of ZnPc is almost identical to

a)   b)   c)

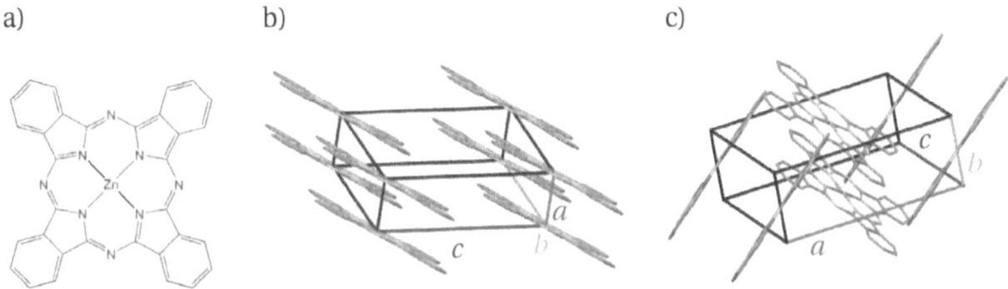

Figure 3.2.: a) Molecular structure of zinc phthalocyanine (ZnPc). The crystal structures of the α [Erk04] and the β polymorph [Luc+20] are shown in b) and c), respectively. The unit cell axes a, b and c are colored red, green, and blue, respectively. Visualization produced by *Mercury 4* [Mac+20] from *cif*-files of the respective publication.

the well studied copper-Pc (CuPc) whose structural data is commonly used as a reference [Bru+10]. Recently, the ZnPc β phase single crystal structure has been disclosed [Luc+20] and proven to be similar to the CuPc β phase [Bro68]. The unit cells of the α and the β phase[2] are shown in figure 3.2 b) and c), respectively, and the unit cell parameters are listed in table 3.1.

	α-phase [Erk04]	β-phase [Luc+20]
Space group	$P\bar{1}$ (2)	$P\,2_1/n$ (14)
a	3.8052 Å	14.5347(6) Å
b	12.9590 Å	4.8529(2) Å
c	12.0430 Å	17.1927(7) Å
α	90.6400°	90°
β	95.2600°	106.201(2)°
γ	90.1200°	90°
V	591.267 Å^3	1164.54 Å^3
Z	1	2

Table 3.1.: Crystal structure comparison of the ZnPc α [Erk04] and β polymorph [Luc+20]. Comparison comprises space group (short name and number), cell length (a, b, and c), cell angles (α, β, and γ), cell volume V, and molecules per unit cell Z.

The *Q-band* absorption, *i.e.* the $S_0 \rightarrow S_1$ transition, usually consists of two absorption lines. For thin films the transitions have been reported between 1.72 eV to 1.76 eV and

---

[2]As suggested in literature [Bru+10], it is assumed that the ZnPc α phase is identical to CuPc α phase [Erk04] shown here.

1.97 eV to 2.0 eV [He+15][Bre+15] and for measurements in solvents or in the gas phase values from 1.85 eV to 1.88 eV and from 2.05 eV to 2.07 eV [EG70][MS95][The+15] are published. The origin of the doublet was long ascribed to a $\pi \rightarrow \pi^*$ transition and an alleged $n \rightarrow \pi^*$ transition from the *aza*-system to the first excited state of the ligand as described by Mack and Stillman [MS95] based on the theoretical description developed for porphyrins by Gouterman [Gou61] and is still used today to describe transitions in ZnPc thin films [He+15]. However, new calculations suggest that the involvement of a symmetry forbidden $n \rightarrow \pi^*$ transition is not needed to describe the transition doublet, but assuming a symmetry breaking that lifts the excited state degeneracy suffices for an excellent agreement between theory and experiment [The+15].

3.1.3. Electrode and Transport Materials

The fabrication of OLEDs requires suitable electrode materials for charge injection as well as materials enabling efficient charge transport to the active recombination zone. The electrode as well as the transport materials used in this work will be briefly introduced. The work functions or HOMO and LUMO energies are shown in the energy diagram in figure 3.3.

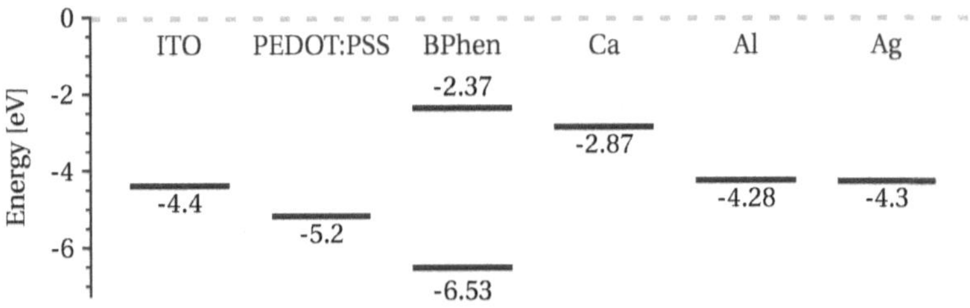

Figure 3.3.: Energy diagram of electrode and transport materials labeled by the respective abbreviations or element symbols. From left to right: ITO, PEDOT:PSS, BPhen, Ca, Al, Ag

Indium tin oxide (ITO) is a heavily tin oxide (SnO_2) doped indium oxide (In_2O_3) semiconductor [Kim+99] and is widely applied as a transparent electrode in OLEDs [Cao+14]. Usually the resistivity is $2 \cdot 10^{-4}\,\Omega\,cm$ to $4 \cdot 10^{-4}\,\Omega\,cm$ at a transmittance of about 90 % in the visible range due to the large band gap of 3.8 eV to 4.2 eV [Kim+99]. However, the optical as well as the electrical properties strongly depend on the processing conditions [DJ84]. The work function determined by photoelectron spectroscopy

3. Materials and Methods

amounts to -4.4 eV to -4.5 eV [Par+96], but depending on the post fabrication treatment values down to -5.2 eV have been achieved [Mas+99][Din+00]. The ITO/glass substrates used in this work were fabricated by Dr. S. Höhla, Institute for Large Area Microelectronics, University of Stuttgart. Their sheet resistance was previously determined to $14.2\,\Omega/\square$ [Ste15].

Silver (Ag) is a 3d transition metal with a very low resistivity of $1.59\cdot10^{-6}\,\Omega\,\mathrm{cm}$ (conductivity $6.3\cdot10^5\,\mathrm{S\,cm^{-1}}$) [Man+99]. The melting point of $(960\pm1)\,^\circ\mathrm{C}$ [Gar+04] enables thermal deposition via vacuum evaporation from tungsten boats. The work function depends on the respective crystal surface [Ish+99]. For the (110), (100) and (111) single crystal surfaces the work functions are $\Phi_{110} = (-4.14\pm0.04)\,\mathrm{eV}$, $\Phi_{100} = (-4.22\pm0.04)\,\mathrm{eV}$, and $\Phi_{111} = (-4.46\pm0.04)\,\mathrm{eV}$ [CM82]. As the silver contacts usually grow polycrystalline, a reference value for the work function is given by $\Phi_{\mathrm{poly}} = (-4.30\pm0.02)\,\mathrm{eV}$ [HR64].

Aluminium (Al) and Calcium (Ca) are usually used in a combination as a cathode material. Both have a high room temperature conductivity ($\sigma_{\mathrm{Al}} = 3.7\cdot10^5\,\mathrm{S\,cm^{-1}}$ and $\sigma_{Ca} = 2.9\cdot10^5\,\mathrm{S\,cm^{-1}}$ [Hay14]) and melting points ($T_{\mathrm{m,Al}} = 660\,^\circ\mathrm{C}$ and $T_{\mathrm{m,Ca}} = 842\,^\circ\mathrm{C}$ [Hay14]) enabling evaporation from tungsten boats, although Al reacts with tungsten [CS31] the limited solubility makes evaporation still possible [Str33]. The work function of aluminium is rather high, with values of $\Phi_{110} = (-4.06\pm0.03)\,\mathrm{eV}$, $\Phi_{110} = (-4.20\pm0.03)\,\mathrm{eV}$ and $\Phi_{111} = (-4.26\pm0.03)\,\mathrm{eV}$ for different crystal facets and $\Phi_{\mathrm{poly}} = (-4.28\pm0.01)\,\mathrm{eV}$ for polcrystalline films [EM73], which can induce high electron injection barriers for most organic semiconductors. Calcium on the other hand has a work function of $\Phi_{\mathrm{poly}} = (-2.87\pm0.06)\,\mathrm{eV}$ for polycrystalline films [GR71], thus being suitable for electron injection, but is prone to fast oxidation and quickly forms the insulator $Ca(OH)_2$ in the presence of water in the environment. To make use of its low work function it is usually evaporated as a thin layer ($\approx 10\,\mathrm{nm}$) and capped with a thick aluminium layer ($\approx 150\,\mathrm{nm}$) to prevent oxidation. However, in most cases, an additional encapsulation is still needed.

Poly(3,4-ethylenedioxythiophene) polysterene sulfonate (PEDOT:PSS) is one of the most frequently used hole injection and transport materials in organic electronics. It comprises the highly conductive polymer poly(3,4-ethylenedioxythiophene) (PEDOT), whose low solubility is compensated by addition of the water-soluble polysterne sulfonate (PSS) during polymerization yielding a PEDOT:PSS water composite with a conductivity of $\approx10\,\mathrm{S\,cm^{-1}}$ [Hua+05]. Functionalization of an underlying ITO electrode with PEDOT:PSS lowers the hole injection barrier into the consecutive organic layer [CBA01] and smoothens the polycrystalline ITO film surface, thereby reducing electrical shorts across the subsequent organic layers [PYF03]. The utilized PEDOT:PSS is the commercially available aqueous dispersion CLEVIOS™P VP AI 4083 purchased from *Heraeus* and can be processed by spin coating. According to the data sheet [Her19] dried films have a work function of $\Phi_{\text{PEDOT:PSS}} = 5.2\,\text{eV}$ and a resistivity of $500\,\Omega\,\text{cm}$ to $5000\,\Omega\,\text{cm}$.

Bathophenanthroline (BPhen)   or 4,7-diphenyl-1,10-phenanthroline is commonly sandwiched between the active layer of the device and the metal contacts and acts as an electron transport material, exciton blocking layer, and penetration barrier for thermally deposited metal atoms [Ste+12]. The HOMO and LUMO energies determined by UPS and IPES are given as $-6.53\,\text{eV}$ and $-2.37\,\text{eV}$ [Mey+10] and electron mobilities as high as $10^{-4}\,\text{cm}^2\,\text{V}^{-1}\,\text{s}^{-1}$ have been observed in time-of-flight experiments [Nak+00]. The lack of absorption in the visible range [Wu+10][Kum+08], making BPhen not susceptible to resonant energy transfer for commonly used emitters, and the easy processability via vacuum sublimation lead to the widespread technological application of BPhen.

3.2. Sample Preparation

Sample preparation marks a crucial first step in tackling a research question, as the parameters and methods chosen determine the sample's macroscopic and microscopic properties thus, *vice versa*, influencing molecular interactions which are crucial for *e.g.* excited state dynamics as discussed in section 2.3.4. Therefore, the methods used for crystal growth as well as for molecular thin film and device fabrication will be introduced in the consecutive sections.

3.2.1. Crystal Growth: Horizontal Vapor Deposition

There are various techniques and setups for the preparation of organic semiconductor single crystals from the vapor phase, for example the Lipsett method [Lip57], and they are roughly divided into open and closed system based [KSP10]. The technique used in this work is the *horizontal physical vapor deposition*, first implemented by Laudise *et al.* [Lau+98], using an open system setup with an inert gas flow. The setup's schematic and the working principle is explained on the basis of figure 3.4.

The furnace comprises an outer fused silica glass tube wrapped with a resistance wire. The middle of the coil is contacted and set to a common ground potential while

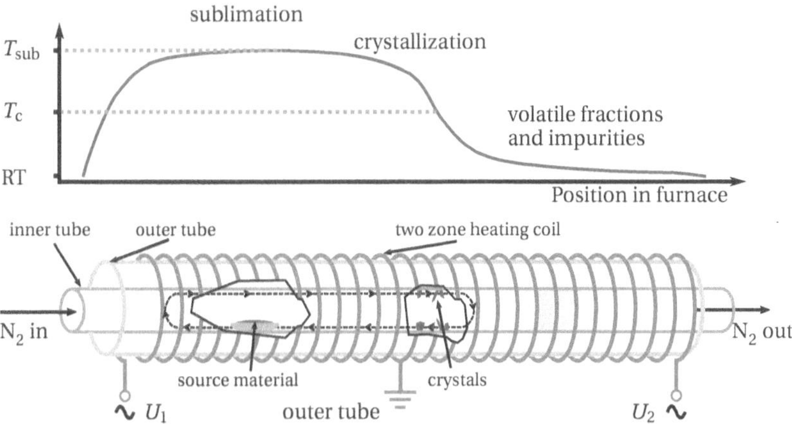

Figure 3.4.: Crystal growth by horizontal physical vapor deposition. At the bottom the horizontal furnace is shown while the top half displays the temperature profile along the tube generated by the heating coil wrapped around the outer tube. The source material in the inner tube is sublimed at a sufficiently high temperature and transported by the inert N_2 gas stream to a zone of lower temperature where it crystallizes. Potential lighter impurities deposit at the cold end of the inner tube. Adapted from [Lau+98].

at both ends the altering potentials U_1 and U_2 are applied. Applying voltages $U_1 > U_2$ leads to two temperature zones inside the furnace with a steep gradient in between as shown in figure 3.4. An inner borosilicate glass tube, trade name DURAN^{TM}, is centered inside the outer tube. The source material is either placed in a commercially available silica glass combustion boat or a tray folded from cleaned aluminium foil inside the inner tube. The position is carefully chosen for a constant temperature over the entire length of the boat. An inert gas flow[3] of 30 sccm to 50 sccm of molecular nitrogen N_2 with 6N purity is applied as inert transport gas. The maximum gas velocity at the position of the source material $x = 30\,\mathrm{cm}$ of the N_2 jet can be estimated to

$$u_{max} = \frac{1}{2\pi^2} \frac{Q^2}{v a^2} \frac{1}{x} \approx 1\,\frac{\mathrm{mm}}{\mathrm{s}} \tag{3.1}$$

at a flow rate of $Q = 30\,\mathrm{sccm}$ and with $v = 0.159\,\mathrm{cm}^2\,\mathrm{s}^{-1}$ as the kinematic viscosity [Hay14] and $a = 2\,\mathrm{mm}$ as the radius of the gas inlet tube [Lau+98]. At these conditions, the stream is superimposed by buoyancy driven convection (*c.f.* figure 3.4) as a result of the steep temperature gradient with peak velocities, estimated for a setup with similar

---

[3]The commonly used unit to measure the mass flow rate of a gas is *standard cubic centimeters per minute* $1\,\mathrm{sccm} := 1\,\mathrm{cm}^3\,\mathrm{min}^{-1}$ [BGV14].

dimension, of $12\,\mathrm{cm\,s^{-1}}$ [Lau+98]. However, gas flow simulations of buoyancy driven convection in closed system based crystal growth have shown that oversimplifications of the flow mechanics have to be treated with care [Ros+97].

The source material is heated to, or slighty above, its sublimation temperature T_{sub} and the molecules in their vapor phase are transported along the tube. As the temperature drops below a critical point T_c the vapor supersaturates and nucleation at the inner tube's wall starts followed by crystal growth as described in 2.2.3.

The growth process is stopped by slowly cooling down the tube to room temperature by automatically reducing U_1 over the course of 8 h to 18 h while keeping U_2 constant to avoid fluctuations in the ambient temperature causing disturbances. This procedure does change the temperature gradient which can alter the growth parameters at the beginning of the cooling period as material is still subliming. However, a slow cooling is mandatory to reduce thermal stress on the crystal as some materials experience highly anisotropic thermal contraction upon cooling and even expansion for some direction [UWD43][Haa+07]. After cooling, the crystals are retrieved by carefully cutting the inner tube close to the growth zone.

A benefit of the horizontal physical vapor deposition, besides the need for only small amounts of source material ($\approx 30\,\mathrm{mg}$), is the purification of the source material during the growth process [SW07][Lau+98]. Impurities with a high sublimation temperature will reside in the combustion boat, while volatile impurities with a much lower (re)sublimation temperature will be transported by the non-convective part of the transport gas flow to the colder end of the inner tube and deposit as a film on the tube's wall.

To maximize this benefit of purity the inner tube and the glass combustion boats are subjected to a cleaning process. The tubes are rinsed three times with acetone (*p.a.* grade) followed by deionized water. After that, they are rinsed with isopropyl alcohol(*p.a.* grade), and blow dried with N_2 (5N purity). The silica glass combustion boats are cleaned in acetone and isopropyl alcohol (both *p.a.* grade) in an ultrasonic bath for ten minutes each, rinsed with deionized water in between, and blow dried with N_2 (5N purity). Self made aluminium boats are folded from thoroughly acetone and isopropyl alcohol cleaned ROTILABO® aluminium foil (thickness $30\,\mu\mathrm{m}$). The combustion boat is placed in the inner tube and the furnace is heated above T_{sub} with an N_2 flow of $200\,\mathrm{sccm}$ for $\approx 12\,\mathrm{h}$ in both parts of the furnace to eliminate any residual impurities on the boat and on the tube wall which might contaminate the growing crystals later on. If a silica glass combustion boat has to be reused for a different material it is treated beforehand with a $0.02\,\mathrm{M}$ solution of potassium permanganate and the resulting manganite is solved with concentrated (37 %) hydrochloric acid.

3.2.2. Thin Film and Device Preparation

For the preparation of thin films and metal contact structures, different techniques were used, dependent on the type of material.

Small molecules   were deposited via *physical vapor deposition* in high vacuum with a base pressure better than 10^{-8} mbar. For this purpose the substrate is mounted upside down on a heatable or coolable copper block approximately 30 cm above the evaporation sources. The material is evaporated from a *Knudsen cell* comprising a boron nitrite crucible wrapped by a tungsten wire as resistance heater. The respective evaporation temperature can be monitored via a typ K thermocouple welded to a tungsten ring positioned at the cell's aperture. Sublimed from the Knudsen cell, the molecules enter the vacuum chamber. The mean free path length $\langle l \rangle$ for a given ambient pressure P and temperature T can be approximated by the kinetic gas theory to [MS97]

$$\langle l \rangle = \frac{k_\mathrm{B} T}{\sqrt{2} \sigma P} \tag{3.2}$$

with σ as the molecules' collision cross section. Assuming the molecules' cross as a circular area of 5 Å radius, $\langle l \rangle$ can be estimated to $\langle l \rangle \gtrsim 10^3$ m for a base pressure of $P = 10^{-8}$ mbar and room temperature[4], *i.e.* $T = 300$ K. This is, of course, much larger than the dimensions of the vacuum chamber and the distance to the substrate, thus, justifying the common assumption of a ballistic transport between source and substrate. The film thickness and, hence, the deposition rate is monitored via a quartz crystal microbalance positioned next to the substrate [Sau59].

PEDOT:PSS   hole transport layers are deposited on pre-cleaned ITO substrates via *spin coating*. The aqueous PEDOT:PSS is applied so that it wets the entire substrate surface. The substrate is fixed via a vacuum chuck on a rotational plate. The fast spinning of the plate skids the solution from the center of the substrate to and off the edges while simultaneously the solvent evaporates leaving behind a solid polymer film covering the whole substrate. Its final thickness is dependent on various parameters such as the rotation speed and duration of spin coating [SD03]. As PEDOT:PSS is highly susceptible to ambient water which reduces its conductivity [Nar+08], the films are annealed on a hot plate for 20 min at 120 °C after spin coating and prior to further processing. After the annealing process they are immediately transferred into the intended vacuum system for the next preparation steps. The films were prepared at 3000 rounds per minute

---

[4]Of course, the gas molecules will have a higher temperature when leaving the crucible. Even though this is a lower limit for the mean free path length, the usual sublimation temperature of organic molecules is in the same order of magnitude.

for a period of 60 s with a ramping time of 5 s yielding a film thickness of approximately 50 nm [Bre16].

Metal Contacts are prepared by metal evaporation under high vacuum utilizing a shadow mask to define the contact structure. Silver contacts of a total thickness of ≈ 70 nm were deposited by evaporation in high vacuum (base pressure $\leq 5 \cdot 10^{-7}$ mbar) from a tungsten boat at a rate of $12\,\text{Å s}^{-1}$. Calcium/Aluminium contacts are prepared in a vacuum chamber integrated in a JACOMEX glovebox system operating under an inert N_2 atmosphere to minimize Ca oxidation. Evaporation of 10 nm Ca at $0.2\,\text{Å s}^{-1}$ followed by 150 nm Al at $2\,\text{Å s}^{-1}$ to $5\,\text{Å s}^{-1}$, both from tungsten boats, are carried-out at a base pressure of $\leq 10^{-6}$ mbar. To avoid alloy formation with tungsten, the boat for Al deposition has to be renewed for each evaporation cycle. The contact structure is encapsulated *in-situ* in the glove box with a glass cover slip fixed with a commercially available, low vapor pressure epoxy (trade name TORR SEAL®) to prevent Ca oxidation after transferring the samples out of the glovebox system for further characterization.

3.3. Structural Characterization

The characterization of the microscopic structure of a sample constitutes an important part in the examination and understanding of optoelectronic phenomena. In this work, atomic force microscopy and X-ray diffraction have been used as complementary methods to characterize the surface morphology as well as to get insights in the bulk crystallinity and order.

3.3.1. Atomic Force Microscopy

Atomic force microscopy (AFM[5]) is a highly sensitive technique to analyze the morphology of a surface with close to atomic resolution and has been developed by Gerd Binning, Christoph Gerber and Calvin Quate in 1986 [BQG86] [RH90].

The working principle is straight forward and relies on the forces between atoms and molecules on atomic length scales, in particular the attractive van der Waals force and the Pauli repulsion (*cf.* 2.1.3) and is explained, *e.g.* in the review paper of Meyer [Mey92] and standard textbooks on surface analysis [Leg09] [Voi15], which were used to compile the description of the measurement technique presented in this section. A microscopic tip fixed on a cantilever is moved over the surface while the cantilever's deflection is

---

[5]The abbreviation is used further on in the text to refer to the technique (microscopy) as well as to the apparatus (microscope).

measured and is fed back into a height control feedback loop. From the cantilever displacement the height profile of the surface can be reconstructed.

The AFM used in this work is a Veeco Dimemsion Icon which measures the cantilever displacement by beam deflection. In this method the back side of a reflective cantilever is illuminated by a laser beam. The reflection is directed on a four quadrant photo diode and the difference in intensity on the upper and lower diode's half measures the displacement. If the measurement is conducted in the *static contact mode* the tip is always "in contact" with the sample surface, whereas contact of tip and surface means that the tip is forced to such close proximity to the surface by the cantilever that a constant repulsive force pushes the tip away from the surface which, *vice verse*, bends the cantilever. Knowing the cantilever's elastic properties its displacement is a direct measure for the sample topography. However, this technique continuously stresses the surface by a mechanical force. This is why usually a dynamic intermittent mode, the so-called *tapping mode*, is used in many of today's measurements. An oscillating tip, driven near its resonance frequency, is kept in 10 nm to 100 nm distance apart from the sample surface. As the distance between the tip and the surfaces changes, the tip traverses the attractive as well as the repulsive range of the interaction potential (*c.f.* figure 2.5) and hence, the system can be described as a driven anharmonic oscillator. The system's resonance frequency as well as the damping of the amplitude strongly depends on the distance between the tip's equilibrium position and sample surface. The oscillating tip moves across the surface and the feedback loop maintains either the chosen frequency or amplitude by compensating the damping via adjusting the height of the cantilever and hence, mapping the surface topography. At the lower turning point of each oscillation cycle the tip surface interaction includes energy dissipation from the driven oscillation to the sample surface which directly influences the phase between the driving and the cantilever oscillation. This additional *phase* shift is directly related to the amount of dissipated energy to the sample which, *inter alia*, depends on the surface material. Thus, for heterogeneous samples, mapping the oscillation's phase in addition to the surface topography yields insights into the surface distribution of the different constituents .

The Veeco Dimension Icon AFM was used together with OLYMPUS OMCL-AC160TS silicon tips [Oly18] operating at a resonance frequency in the range of 200 kHz to 400 kHz and a typical force constant of 26 N m^{-1}. The data was analyzed with the open source software Gwyddion [NK12].

3.3.2. X-Ray Diffraction and Reflectometry

X-ray scattering provides a mostly non invasive tool for determining integral structural parameters such as the crystal structure and film thickness as well as for providing insights in statistical microscopic parameters such as the mosaicity, average crystal height, and surface roughness. The non invasive character has to be treated with care depending on the property it refers to and the sample material. For example, in organic semiconductors, X-ray exposure can introduce deep traps for charge carriers by breaking molecular bonds and creating new chemical species, and hence, strongly influences electronic transport properties [Pod+04].

All measurements were performed on a General Electrics XRD 3003 TT thin film reflectometer operating with a monochromatic Cu $K_{\alpha 1}$ source with $\lambda = 1.5406$Å. The system utilizes the Bragg-Brentano geometry, *i.e.* the horizontally fixed sample is orbited by the X-ray source and the detector mounted on independent goniometers moving around the sample on the same circle (*Rowland circle*) within one plane.

X-ray diffraction (XRD) experiments are described by means of the kinematic diffraction theory and the main results of this will be presented in the following. The theoretical basics are treated in many solid state physics monographs [Dem10] and will not be discussed here.

If a monochromatic X-ray beam of wavelength λ hits a crystal surface with Miller indices (hkl) at an incident angle θ, a part of it will be reflected by the surface atoms following *Snell's law*, while the remaining photons will penetrate into the crystal. At the consecutive lattice plane, again, some photons will be reflected by the atoms while the rest will penetrate deeper into the crystal. The two resulting reflected beams now have a phase difference due to the difference in their respective path length and interfere destructively, unless the incident angle and the lattice plane distance d_{hkl} are related by the following relation known as *Bragg's law*

$$2d_{hkl}\sin\theta = n\lambda. \tag{3.3}$$

Here, n is an interger number marking the order of the resulting intensity peak [Dem10]. Rewriting the Bragg equation (3.3) in terms of the momentum transfer $\vec{q}$ along the surface normal, with $q = |\vec{q}| = 4\pi/\lambda \sin(\theta)$, yields q as a multiple of the respective reciprocal lattice distance

$$q = n\frac{2\pi}{d_{hkl}}. \tag{3.4}$$

For an idealized, infinitely thick single crystal the line width of an intensity peak in θ-2θ geometry, *i.e.* in the case of source and detector enclosing the same angle with the sample, is only limited by the beam divergence and the detector slits. In contrast, for a sample with a finite, small number of lattice planes N in the crystal direction [hkl], the intensity $I(q)$ is determined by the *Laue interference function* [Jam62]

$$I(q) \propto \frac{\sin\left(\frac{N q\, d_{hkl}}{2}\right)^2}{\sin\left(\frac{q\, d_{hkl}}{2}\right)^2} \tag{3.5}$$

which possesses a main intensity maximum for q-values satisfying the Bragg condition (3.4) accompanied by $N-2$ subsidiary maxima at $q = \pm(2k+1)\pi/N$ with $k = 1, \dots, N-2$ [Dür02]. Normalizing the Laue function by the number of lattice plains yields the contribution of each plane to the overall signal. The signal resembles the N-th *Fejér kernel* forming a *Dirac sequence* and, hence, converges to a δ-function for $N \to \infty$ [Köh62] as expected for an infinitely thick crystal.

As evident by the previous description, the crystallite height h can be determined as $h = d_{hkl}N$ by extracting N from the distance of two adjacent subsidiary maxima. If for higher values of N the subsidiary maxima become harder to resolve or if structural disorder suppresses defined Laue oscillations, the *Scherrer equation* can be used to estimate the crystallite height [Gui94]

$$h = \frac{0.9\lambda}{\cos(\theta)\,\Delta(2\theta)} \tag{3.6}$$

where θ is half of the diffraction angle 2θ of the corresponding (hkl)-reflex and $\Delta(2\theta)$ is the *full width half maximum* (FWHM) with respect to the 2θ abscissa. For a complete consideration the instrumental broadening Δ_b has to be taken into account. Thus, for a measured FWHM $\Delta_\mathrm{m}(2\theta)$ the Scherrer equation has to be modified by $\Delta(2\theta) = \sqrt{\Delta_\mathrm{m}(2\theta)^2 - \Delta_\mathrm{b}^2}$. As stated by Warren [War41], this correction can be neglected as long as $\Delta_\mathrm{b} < {}^1/_5\Delta_\mathrm{m}$.

In a *rocking scan* the source and detector are kept at constant angles fulfilling the Bragg condition stated by equation (3.3) for a dedicated peak. They are then rotated around the sample and thus, probing its *mosaicity, i.e.* the angular tilting of its individual grains with respect to the surface normal. A sharp peak indicates highly oriented crystallites while a broadening of the signal, *i.e.* its rocking width, is caused by tilted domains. The FWHM of the specular rocking peak on a (hkl) Bragg reflection is the sum of various contributions and can be written as [Aye94] [Spi+09]

$$\mathrm{FWHM}_{hkl}^2 = \underbrace{\beta_0^2 + \beta_\mathrm{i}^2}_{\text{inital}} + \beta_\mathrm{screw}^2 + \beta_\mathrm{strain}^2 + \beta_\mathrm{dim}^2 + \beta_\mathrm{curve}^2. \tag{3.7}$$

The first two terms relate to the intrinsic width β_0 for an ideal crystal and the width of the incidental beam β_i, *e.g.* resulting from the monochromator used. At screw dislocations, *i.e.* imperfections in the crystal lattice, usually a tilting of the crystallites is induced which is accounted for by β_{screw} and strain induced broadening surrounding these dislocations is described by β_{strain}. Both are dependent on the dislocation density [Aye94]. The last two terms concern broadening due to the size of the crystallites β_{dim} and due to sample curvature β_{curve}, *i.e.* substrate curvature in thin film samples or curvature of a single crystal specimen. With a rocking measurement an estimation of the dislocation density is possible, either from the evaluation of the specular FWHM or by comparison of the specular intensity to the off specular FWHM as described in [Nic+04].

X-ray reflectometry (XRR) is performed at very low incident angles up to approximately 5° and allows determination of the overall film thickness of a thin film sample between 3 nm to 200 nm depending on the angular resolution, and even more important, on the sample's interface roughness as described below, mainly based on [Spi+09]. At grazing incidence, *i.e.* almost parallel incidence of the beam with respect to the sample surface, the beam is totally reflected at the surface as the refraction index for X-rays is below unity for almost all materials. After exceeding the *critical angle*, which depends on the electron density of the material, a fraction of the beam is transmitted into the substrate and the intensity $I(q)$ quickly decreases with increasing momentum transfer according to the *Fresnel law*. For a smooth surface the intensity decrease in relation to the incident intensity I_0 can be described by [CBB14]

$$\frac{I(q)}{I_0} \propto q^{-4}. \tag{3.8}$$

A real surfaces is at least atomically rough and the roughness can be characterized by the *root mean square* (rms) of the height deviations σ_{rms}. This leads to a second factor in the intensity decline

$$\frac{I(q)}{I_0} \propto q^{-4} e^{-2q^2 \sigma_{rms}^2} \tag{3.9}$$

giving rise to the possibility of a non contact determination of an integral surface topography measure [CBB14]. If the sample comprises at least two layers of materials with a different refraction index, *e.g.* an organic thin film on an inorganic substrate, after exceeding the critical angle, the X-ray beam can enter the cover layer. A part of the beam is reflected at the cover layer's surface and a part of the transmitted beam is reflected on the shared interface with the subsequent layer leaving the sample parallel to the already reflected beam. The interference of these two beams leads to so called *Kiessig fringes* in the reflection intensity as a function of momentum transfer. The distance between two adjacent maxima or minima of momentum transfer $q_n > q_{n-1}$ is related to the layer thickness t by

$$t = \frac{2}{q_n - q_{n-1}}.$$ (3.10)

Today, usually the *Parrat-formalism* is applied [Par54] to determine film thicknesses and interface roughness from small angle XRR diffractograms. It is based on the Fresnel law and yields a recursion formula for the reflected intensity of a multi layer system in which also the surface roughness and the interface roughness between layers is considered.

3.4. Optical and Electronic Characterization

The main techniques to characterize electronic excited states used in this work are based on measuring the photon absorption and emission of the various samples of interest, by means of steady state absorption and photoluminescence as well as differential reflectance spectroscopy. Prior to describing the individual experimental setups used in this work, the following section will discuss the necessary data handling to correctly interpret the obtained results.

3.4.1. Relating Absorption and Emission Spectra

A light beam of incidental intensity I_0 passing through a non reflecting sample with n' molecules per cm^3, each having the wavelength dependent photon absorption cross section $\sigma'(\lambda)$ in cm^2, will be attenuated at a distance x according to the *Lambert-Beer law* [KB15]

$$I(x) = I_0 e^{-\sigma'(\lambda)n'x}.$$ (3.11)

For thin film samples the *absorption coefficient* $\alpha(\lambda) = \sigma'(\lambda)n'$ with $[\alpha] = 1\,\mathrm{cm}^{-1}$ is used. For solutions with molar concentration c_M in $\mathrm{mol\,l}^{-1}$, the exponent is rewritten as $\epsilon\,c_M x$ with $\epsilon = 10^{-3}\sigma' N_A$ as the *molar extinction coefficient*[6] with $[\epsilon] = 1\,\mathrm{l\,mol}^{-1}\mathrm{cm}^{-1}$ [KB15]. N_A is the Avogadro constant.

In section 2.3.1 it was shown that the absorption rate depends on the strength of the surrounding light field and the Einstein coefficient $B_{12} \propto |\vec{M}|^2$ which, in return, depends on the transition dipole moment $\vec{M}$ mediating the electronic transition. As described in [AGR06] and [MK13], constituting the main references for this section,

---

[6]If the Lambert-Beer law is written in terms of a decadic logarithm the *decadic extinction coefficient* $\epsilon_d = \ln(10)\epsilon$ is used [Par07].

the measured quantities α and ϵ for any transition $1 \rightarrow 2$ at energy E are related to the Einstein coefficient and hence the dipole moment via

$$\alpha, \epsilon \propto E \cdot B_{12} \propto \left| \vec{M}_{12} \right|^2 . \tag{3.12}$$

This implies that for a quantitative analysis of an absorption spectrum, the experimentally determined quantity has to be divided by the related transition energy E. This is shown for a generic molecular absorption spectrum explicitly considering the Franck-Condon vibronic progression in figure 3.5 (blue). The blue dotted line, representing the experimentally determined absorption, deviates from the transition spectrum given by equation (2.41) and shown as the solid line.

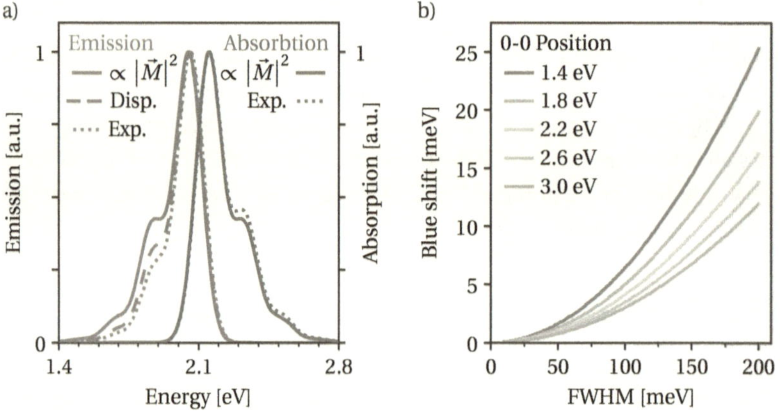

Figure 3.5.: a) Simulated absorption spectrum (blue) and emission spectrum (red) taking into account Franck-Condon vibronic progression. Experimentally obtained spectra are shown as a dotted, transition spectra as solid lines. The Jacobi transformed emission spectrum is depicted as a dashed line. b) Simulation of the blueshift of the alleged 0–0 transition between an as obtained emission spectrum and the actual transition spectrum as function of the FWHM for different central energies of the underlying gaussian line shape. Adapted from [AGR06].

The energy dependent intensity distribution of emitted radiation is usually described in terms of spectral densities, *i.e.* an energy or power density per unit area and/or solid angle **per unit frequency** [Dem11]. Integrating over specific variables of the respective optical experiment, for example the emission area, the solid angle, and observation time, the intensity is described in terms of a *spectral density* $S(\omega)$ and the intensity in the interval $[\omega; \omega + d\omega]$ is given as $|S(\omega)d\omega|$. Spectral densities are usually recorded over a linear wavelength scale[7], hence, the recorded intensity is given as $|S(\lambda)\mathrm{d}\lambda|$. To transform the signal to an energy dependent quantity a coordinate transformation, depending on the underlying dispersion relation, has to be performed [Hea03]. For the photon energy $E = hc/\lambda$ the *Jacobi transformation*[8] yields [MK13]

$$|S(E)\mathrm{d}E| = |S(\lambda)\mathrm{d}\lambda| \quad \Rightarrow \quad S(E) = S(\lambda)\frac{\mathrm{d}\lambda}{\mathrm{d}E} = -S(\lambda)\frac{hc}{E^2}. \tag{3.13}$$

The first expression originates by the conservation of energy and implies that the emitted energy in the wavelength interval $[\lambda; \lambda + \mathrm{d}\lambda]$ is same in the corresponding energy interval $[E; E + \mathrm{d}E]$. The minus sign just notes the reversed direction of the integration and can be dropped. The radiative transition rate $2 \rightarrow 1$ of an excited molecule is determined by the respective Einstein coefficient A_{21} for spontaneous emission which relates the spectral density to the transition dipole moment according to (2.34a) as follows

$$S(E) \propto A_{21} \propto E^3 \left|\vec{M}_{21}\right|^2. \tag{3.14}$$

To relate an experimentally obtained spectrum to the transition dipole moment, it has first to be transformed to the energy scale according to equation (3.13) and then corrected for the cubic energy dependence of the transition rate. This is shown in figure 3.5 a). An experimentally obtained emission spectrum (red dots) transforms into the dashed spectra after the dispersion correction. After eliminating the cubic energy dependency, the intensity distribution is proportional to the transition spectrum shown as the solid line. Only then, the mirror symmetry predicted by the Franck-Condon principle is visible.

The solid lines, representing the emission and absorption signal proportional to the respective squared transition dipoles are sometimes called *spectral line shapes* [MK13] or *reduced spectra/intensity* [SB16]. In the following the quantity will be referred to as *transition spectrum*, to distinguish it from line shape functions used to deconvolute emission and absorption spectra, and will be noted by the symbol $\bar{I}(E)$ in the following.

Due to a lack of rigor in the application of the appropriate corrections of absorption and emission spectra some flawed interpretations of optical data have been published [SL99]. As it becomes evident by figure 3.5 a), transforming the experimental spectrum to the energy scale without the Jacobi transformation and interpreting it as a transition spectrum, leads to a severe underestimation of the related Huang-Rhys parameter. What becomes not as evident by the shown emission data, is that the peak position of

---

[7] This stems from the fact that most spectrometers use diffraction gratings which approximately have a linear dispersion relation as function of the wavelength [Hea03].

[8] The name presumably stems from the Jacobi-Matrix formalism in coordinate transformation [Tim06] which is reduced to a scalar in the one dimensional case.

the experimental data is blue shifted with respect to the transition spectrum. In figure 3.5 b) the blue shift is shown as a function of the FWHM of a single gaussian emission at different energies (0-0 position). It is evident that for broad peaks at low emission energy the blue shift can be in the range of the thermal energy at room temperature indicating that one should proceed with caution when determining transition energies directly from experimentally obtained spectra without the above mentioned corrections.

3.4.2. μ-Photoluminescence Setup

In a *photoluminescence* (PL) experiment a sample is excited by a laser and the resulting photoluminescence is recorded by a detector. Figure 3.6 shows the μ-PL setup used in this work. For the excitation an Oxxius LaserBoxx 532 nm fiber coupled continuous wave laser or a Coherent OBIS LS/LX 685 nm continuous wave laser can be used by inserting an additional mirror (AM). Each laser is operated with a spectrally narrow laser line band pass filter. The 532 nm laser is refocused by a lense system after leaving the optical fiber. The PL signal is measured with a Princeton Instruments Acton SP2500i spectrometer containing three diffraction gratings with $150\,\mathrm{lines\,mm^{-1}}$, $300\,\mathrm{lines\,mm^{-1}}$ and $1500\,\mathrm{lines\,mm^{-1}}$. Spectrally resolved PL is recorded with a PIXIS 100BR_ eXcelon CCD camera. Intensity measurements can be carried out via a LASER COMPONENT COUNT®-100C avalanche photo diode (APD). Below the optical plane, the sample is placed on a cold finger in a CryoVac continuous-flow helium cryostat mounted on an nPoint nPBio300 nanopositioning piezo stage. The accessible temperature ranges from 4 K to 400 K and is controlled by a CryoVac Tic500 temperature controller, while the sample temperature is measured with an OMEGA C670 silicon diode. The nanopositioning stage has a translational range of $300\,\mu\mathrm{m}$ in each spatial direction and is controlled by a nPoint C-300 DSP controller. A computer with a National Instruments DAQ 6323 PCIe card connected with two National Instruments BNC-2110 interfaces enables controlling the nanopositioning stage as well as retrieving the signal from the APD bymeans of an in-house written LabView programm.

Before describing the individual setup configurations for the specific measurements, the setup's operation is presented in a general way. The beam path of the respective excitation laser light is indicated in figure 3.6 by the green line. After leaving the laser source and the subsequent optical elements, *i.e.* the laser line filter and, for the 532 nm laser, the refocusing optics, the light enters a system of two adjustable mirrors, which are used to guide to light through the setup. After passing the subsequent neutral density filters (ND), a fraction of the laser light is reflected onto a Thorlabs PM100D power meter for intensity monitoring and control by a beam splitter (BS1). The second frac-

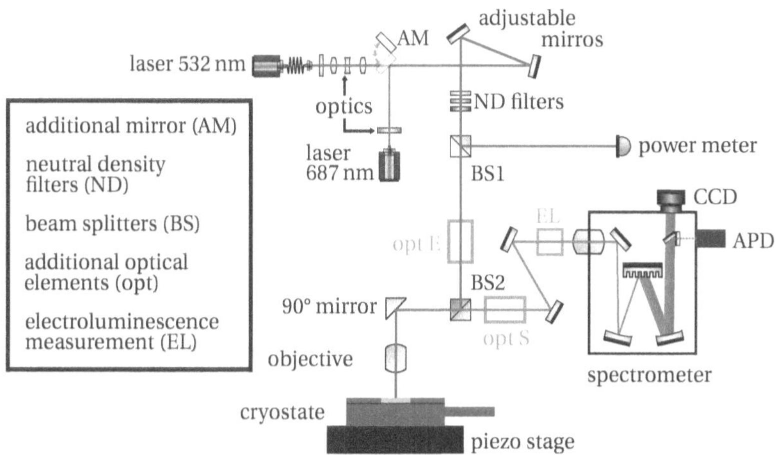

Figure 3.6.: μ-PL setup schematics. Various spatially and spectrally resolved temperature dependent measurements can be performed with this setup. See list of figures for image right attributions.

tion of the excitation light passes additional optical elements (opt E), which depend on the specific application and are described below, before being redirected by a second beam splitter (BS2). The beam is then guided by a SI90° mirror out of the optical plane and into an OLYMPUS SLMPLN 50x objective with a numerical aperture $NA = 0.35$ and a working distance of 18 mm focusing the excitation light onto the sample in the cryostat. The PL signal is collected by the same objective, guided through the beam splitter (BS2) and additional optical elements (opt S), again depending on the application, and finally by two mirrors into the spectrometer. Below, the different usages of the setup with their specific experimental realization are listed briefly:

Spectrally resolved PL studies are performed with either 532 nm or 685 nm laser excitation. For circular polarized excitation a quarter-wave plate preceded by a linear polarizer is added into the beam path at *opt E*. By virtue of the individual laser source's output characteristics, the beam splitters used differ for the two sources. Excitation with 532 nm is performed with a polarizing beam splitter on position *BS1* and an dichroic beam splitter at *BS2* while 685 nm excitation is performed with 90/10 (transmittance/reflectance in %) beam splitters on positions *BS1* and *BS2*. The PL signal is cleared of residual laser light by suitable combination of long pass and notch filters inserted at opt S.

PL intensity mapping is performed via scanning the sample line-by-line with the nanopositioning stage and simultaneously recording the PL intensity at each position by the APD. The maximum spatial resolution is determined by the Rayleigh criterion $0.66\lambda/NA$ [HRW08] with the laser wavelength λ and the numerical aperture NA. This leads to a 1 μm and 1.3 μm lower spatial resolution limit for 532 nm and 685 nm excitation, respectively. The position and the intensity are correlated by LabView, yielding a matrix where each element represents a position on the sample surface and its value the corresponding intensity measured at this spot.

Polarization dependent excitation is performed by adding a linear polarizer and a half-wave plate to the excitation beam path (*opt E*). Rotating the wave plate by an angle α rotates the polarization plane of the linear polarized light by an angle 2α. The respective polarization of the light has to be correlated with the sample orientation in the cryostat.

Polarization dependent emission studies are performed by adding a half-wave plate followed by a linear polarizer to the signal beam path (*opt S*). By rotating the polarization of the emitted light with the half-wave plate, the intensity distribution in the polarization plane can be mapped by projecting a selected polarization with the linear polarizer onto the spectrometer slight. Keeping the linear polarizer fixed ensures that the detected light in the spectrometer always has the same polarization as the diffraction grid's efficiency is polarization dependent. As before, the polarization has to be correlated with the respective sample orientation in the cryostat.

Solvent sample PL can be performed by placing a cuvette at the beam splitter position *BS2* and recording the isotropically emitted light via the two mirror combination in front of the spectrometer. Suitable long pass filters to suppress diffracted laser light can be added at *opt S* if needed. If the respective sample absorption requires it, a 635 nm Thorlabs laser diode can be used on the position of the 685 nm laser to excited the sample.

Electroluminescence (EL) and device characteristics can be carried out with an additional Keysight Technologies 1500A semiconductor parameter analyzer. For EL measurements, the devices are placed in close proximity to the spectrometer aperture (at *EL* in figure 3.6). Due to the low quantum efficiency of the examined devices in this work, spatially resolved measurement in the cryostat by means of the microscope objective were not possible. Recording the plain electronic device characteristics was performed with suitable contact probes.

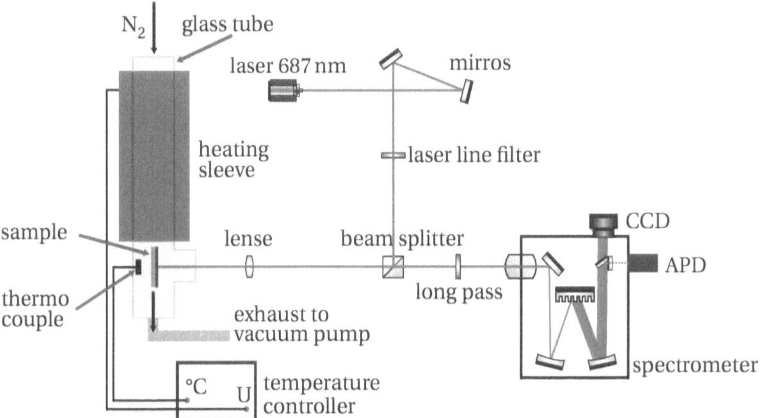

Figure 3.7.: High temperature PL setup. The PL signal of a sample placed in a gas flow oven is recorded over the course of the heating cycle while the temperature is controlled via a feed-back loop. See list of figures for image right attributions.

High temperature PL studies can be performed by placing a gas flow oven in front of the spectrometer as schematically shown in figure 3.7. An approximately one meter long silica glass tube was sheathed by a heating sleeve and thermally insulated with glass wool. At one end, a gas inlet was connected to an N$_2$ (5N purity) gas container, while the other end was connected to a low capacity vacuum pump enabling a constant gas flow of $8\,l\,min^{-1}$. The flowing gas is heated and the temperature at the sample position is measured by a type K thermo couple and maintained by a feed-back coupled Novus Automation Temperature Controller N480D. The 685 nm excitation beam is guided through the laser line filter to a 90/10 beam splitter where it is redirected to a lens of 100 mm focal length, focusing the excitation light onto the sample. The emerging PL signal is collected by the lens and enters the spectrometer aperture after passing a suitable long pass filter to suppress residual laser light. The spectrally resolved PL signal during a thermal cycle can be constantly recorded.

3.4.3. Absorption Measurements

Steady state absorption measurements were performed by a Jasco V-650 UV/VIS spectrometer under ambient conditions. The accessible wavelength range is 190 nm to 900 nm. Either a simultaneous transmission measurement of a sample and a reference sample in double beam geometry is possible or measurement of the absolute transmission and reflectance of a single sample within an integrating Ullbricht sphere is possible by means of different slide in modules for the respective measurement geometry. In this book, the double beam geometry is used for liquid samples while the integrating sphere is used for thin film samples.

Neglecting emittance, the following relation between the transmittance T, the reflectance R, and the absorption A holds

$$1 = T + R + A. \tag{3.15}$$

The quantities T and R refer to the ratio of the transmitted and reflected intensity with respect to the initial intensity and can be used to estimate the spectral absorption using (3.15). For the two beam geometry, the ratio between the transmittance of the sample and its reference yields T.

The absorption or extinction coefficient can be calculated from the transmittance T using Lambert-Beer law (3.11)

$$\alpha = -\frac{\ln(T)}{t} \quad \text{or} \quad \epsilon = -\frac{\ln(T)}{c_M l} \tag{3.16}$$

with t being the film thickness, c_M he molar concentration, and l the optical path length. If the transmittance is determined for a series of film thicknesses t or concentrations c_M a linear fit on a semi-logarithmic scale can be used to determine the respective coefficient. For many thin film samples, the reflectance of the film and the substrate frequently can not be neglected and need to be accounted for. If such a correction is applied in this work, the respective procedure is described in the respective section.

3.4.4. Differential Reflectance Spectroscopy

Differential reflectance spectroscopy (DRS) is a highly sensitive *in-situ* method to probe spectrally resolved optical properties, including absorption, during thin film growth [Pro+05][FF09]. The reflectance of a sample can be studied non invasively by DRS during physical vapor deposition of a film in high vacuum. For this purpose the substrate is illuminated by a white light source and the reflected light is measured by a spectrometer yielding the spectrally resolved reflection signal as function of film thickness $R(t, \lambda)$. The DRS signal for a given film thickness t is defined as the change in reflectance with respect to the plain substrate normalized to the substrate reflection [Bro+11]

$$DRS(t, \lambda) = \frac{R(t, \lambda) - R(0, \lambda)}{R(0, \lambda)} \tag{3.17}$$

and represents a relative change in reflectance as the incident light intensity is eliminated. For an incident and reflected beam parallel to the substrate normal and for film thicknesses much smaller than the respective wavelength $t << \lambda$ the DRS signal is given by

$$DRS(t, \lambda) = -\frac{8\pi n_\mathrm{v} t}{\lambda} \mathrm{Im}\left(\frac{\epsilon_\mathrm{v} - \hat{\epsilon}_\mathrm{f}}{\epsilon_\mathrm{v} - \hat{\epsilon}_\mathrm{s}}\right) \tag{3.18}$$

with $n_\mathrm{v} = \epsilon_\mathrm{v} = 1$ as the refractive index and dielectric function of the vacuum and the complex dielectric functions $\hat{\epsilon}_\mathrm{f}$ and $\hat{\epsilon}_\mathrm{s}$ of the film and the substrate [MA71]. In case of a transparent substrate with an extinction coefficient[9] $\kappa_\mathrm{s} \approx 0$, a refractive index n_s, and $\epsilon_\mathrm{f} = (n_\mathrm{f} - i\kappa_\mathrm{f})^2$ for a non magnetic film [GM12], equation (3.18) simplifies to

$$DRS(t, \lambda) = -\frac{8\pi t}{\lambda}\left(\frac{2 n_\mathrm{f}\kappa_\mathrm{f}}{1 - n_\mathrm{s}^2}\right) \tag{3.19}$$

making the DRS signal a measure for the films's extinction coefficient κ_f [Pro+05]. For opaque substrates, the DRS signal is not as trivial to interpret and has to be analyzed carefully as both, the substrate's and the film's complex dielectric function, determine the signal.

For small changes in the DRS signal, the use of *differential DRS* (DDRS) spectra, *i.e.* the change in spectral reflectance between a film thickness of t and $t + \mathrm{d}t$ normalized to the mean reflectance $\overline{R}$

$$DDRS(t, \lambda) = \frac{\Delta R}{\overline{R}} = \frac{R(t + \mathrm{d}t, \lambda) - R(t, \lambda)}{1/2\,(R(t + \mathrm{d}t, \lambda) + R(t, \lambda))} \tag{3.20}$$

can reveal hidden features that are otherwise not visible in the neat DRS spectra. For very thin films where $\overline{R} \approx R(0)$, the DDRS signal can be calculated as the difference of two subsequent DRS signals $\Delta DRS(t, \lambda) = DRS(t + \mathrm{d}t, \lambda) - DRS(t, \lambda)$. [GWZ15]

The experiments presented and discussed in this book where performed in a collaboration with JProf. Katharina Broch and Clemens Zeiser at the University of Tübingen. All DRS measurements were conducted at normal incidence. The incident light was generated by a DH-2000 deuterium-tungsten halogen light source and the reflection spectra were measured by an Ocean Optics 2000+ UV/NIR spectrometer, together yielding an accessible spectral range between 1.4 eV to 3.0 eV.

---

[9]Here, by the extinction coefficient the imaginary part of the complex refractive index $\hat{n} = n + i\kappa$ is meant. It has to be distinguished from the (molar) extinction coefficient which is historically labeled by the same term.

4. Charge-Transfer States at Pentacene-Perfluoropentacene Interfaces

4.1. Donor-Acceptor Model Interfaces

Donor-acceptor (D-A) interfaces present an important research field driven by fundamental questions and applications alike. Various types of interface charge transfer (CT) states and complexes are proposed and debated as discussed in 2.3.4. CT states are commonly regarded as the intermediate state prior to charge separation in photovoltaic cells, but they can act as a charge carrier traps in ambipolar D-A diodes as well [Opi+09]. A less addressed research question is the influence of the respective molecular orientation at the D-A interface on CT formation and its energy. However, studies suggest suppression of CT formation strongly depends on molecular orientation for planar molecules [Agh+14] as well as a strong influence of the donor orientation on the efficiency of fullerene C_{60} acceptor solar cells due to anistropic charge transfer state formation at the D-A interface [Ran+12][Ran+17].

Figure 4.1.: Scheme of the molecular orientation for a cofacial and a head-to-tail arrangement of PEN and PFP molecules. Adapted with permission from [Ham+20]. Copyright 2020 American Chemical Society.

As discussed previously in section 3.1.1 pentacene (PEN) and perfluoropentacene (PFP) constitute an excellent and intensely studied CT reference system [Bro+17][Kim+19].

The influence of interfacial molecular orientation has been studied in thin film samples via PL [Rin+17], but the results of those studies are difficult to interpret conclusively due to inherent structural defects such as grain boundaries and rotational domains.

This part of the work aims to draw a conclusive picture on the CT interaction in the PEN-PFP model system with respect to two limiting cases of D-A orientation, a cofacial and a head-to-tail arrangement, as sketched in figure 4.1. By means of the optical as well as optoelectronic data, the possible implications for device design and optimization will be examined and employed in prototypical devices. While the cofacial orientation has been extensively investigated in mixed crystalline films, the presented efforts intend to close the knowledge gap by using PEN single crystal templates to create highly ordered PEN-PFP standing stack interfaces with head-to-tail orientation.

4.2. Charge Transfer Complexes in Mixed Films

To establish a reference to the known literature, thin film samples composed of the neat materials as well as a heavily doped film were investigated. Steady state absorption measurements were performed to reproduce the optical absorption properties of the CT complex and complementary structural investigations by means of XRD were conducted.

The PEN, the PFP ,and the PFP doped PEN film were prepared on precleaned glass slides via physical vapor deposition in high vacuum at a nominal thickness of 20 nm. The doped film was prepared as a blend with a volume ration of 4:1 (PEN:PFP) yielding a molecular ratio of 12:1 to preserve the PEN crystal structure while simultaneously introducing plenty cofacially oriented PEN-PFP dimers for a sufficient CT absorption signal in optical measurements.

4.2.1. Structural Investigation

The diffraction patterns of the three samples are shown in figure 4.2. The PEN and blended PEN:PFP film (a) show defined Bragg diffraction peaks at around $0.4\,\text{Å}^{-1}$ and $0.8\,\text{Å}^{-1}$ which can be assigned to (00l) reflexes of PEN. For PFP (b) only a Bragg reflex of low intensity at approximately $0.4\,\text{Å}^{-1}$ is observed.

The crystal height was estimated by two approaches. For the PFP thin film the Scherrer equation (3.6) was used to approximate the crystallite height. For this purpose a Gaussian function with linear background was fitted to the peaks on a 2θ-abscissa (not shown). As the FWHM amounts to $5 \cdot 10^{-3}$ rads, the instrumental broadening, de-

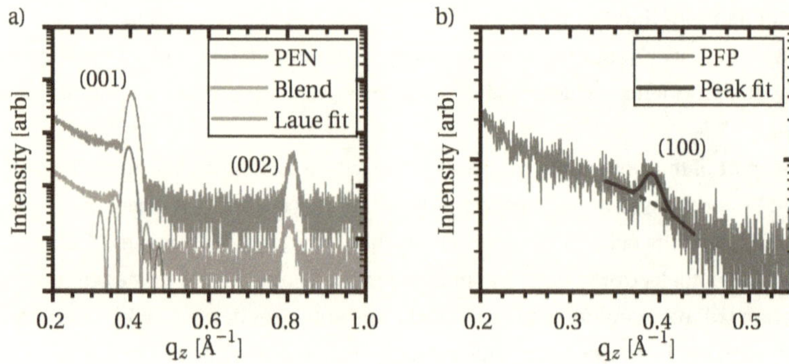

Figure 4.2.: a) XRD pattern of PEN and PEN:PFP mixed films (shifted for clarity) together with an exemplary Laue fit performed with $N = 13$. b) XRD pattern of PFP thin film including the peak fit to determine the PFP lattice constant including a linear background.

termined previously to be $\approx 1 \cdot 10^{-3}$ rad [Han17], has not been considered. As the applied Scherrer equation only roughly approximates the average crystallite height in the limit of thin plates with extended lateral dimensions, the obtained crystallite height of (27 ± 3) nm is in good agreement with the nominal film thickness of 20 nm. For the PEN and the mixed film, the high crystallinity enables fitting of a Laue function with a Fresnel prefactor, given by equations (3.5) and (3.8), to the first order Bragg reflection as shown exemplarily for the mixed film's diffraction pattern in figure 4.2 a) assuming a total of 13 PEN lattice planes. As the PEN film growth is polycrystalline in nature, *i.e.* the film comprises individually growing crystalline PEN islands [Rui+04], the number of lattice planes N has to be interpreted as an average number of lattice planes of the crystallites making up the film. To analyze the data a fit routine using an amplitude factor and the lattice distance d_{001} as free parameters while keeping N at a fixed integer value was employed. The average crystallite height was calculated by $N \cdot d_{001}$ from the best fits. The results are listed in table 4.1. The obtained crystallite heights are in good agreement with the nominal film thickness of 20 nm which indicates that the comprising crystallites extend over the total film thickness. This is beneficial for the later in this work presented implementation in prototypical vertically stacked D-A devices, as grain boundaries along the direction of charge transport are considered one of the most influential hindrances for charge transport and hence a good device performance [Rös+14][Vla+18][Gei+20].

The determination of the out-of-plane lattice constant is sufficient to identify the respective polymorph for neat PEN and PFP in most cases (*c.f.* section 3.1.1). The

	Crystal height [nm]		Laue function		Bragg reflection	
	Scherrer	Laue	N	d_{001} [Å]	q_z [Å^{-1}]	d [Å]
PEN		18.6	12	15.5	0.404	15.6
		20.2	13	15.5	0.812	15.5
Blend		20.5	13	15.8	0.397	15.8
		22.1	14	15.8	0.804	15.6
PFP	27 ± 3				0.932 ± 0.001	16.0 ± 0.4

Table 4.1.: Crystal height and lattice spaccing determined by XRD for PEN, PFP and PEN:PFP blended thin films. Standard errors obtained by fits or calculations smaller than 10^{-3} are not shown.

extracted values of the samples under study are listed in table 4.1 in the column titled "Bragg reflection". The center of the Bragg reflection has been estimated by a Gaussian Fit. In addition, for PFP a linear background correction was applied whereas for the other two samples, the assumption of a constant background was sufficient.

For the neat PEN and the mixed film, the smaller lattice constants obtained from the second order diffractions hint towards a systematic overestimation of the lattice constant due to a *surface displacement error, i.e.* a mismatch of the film surface and the center point of the Rowland circle during the measurement, which is a common systematic error in Bragg-Brentano thin film diffraction experiments [Spi+09]. A systematic error of the extracted lattice constant Δd caused by such a surface displacement error is described by the relation $\Delta d \propto \cos\theta \cot\theta$, and hence, is minimized for larger diffraction angles θ [KP01]. Thus, a linear extrapolation of $\cos\theta \cot\theta \rightarrow 0$ in a $\cos\theta \cot\theta$-d plot yields the correct lattice constant at the intersection of the linear function and the d-axis. For the neat PEN film this extrapolation yields $d = 15.37$Å and comparison with the literature identifies the thin film phase characterized by a 15.4 Å lattice constant [Mat+03b]. Performing the same extrapolation for the blended film, a lattice spacing of $d = 15.43$Å is obtained. As XRD studies on various mixing ratios of PEN and PFP indicate that only for the equimolar mixture a co-crystal is formed and, otherwise, a phase separation of PEN and PFP takes place [Hin+11], the slightly larger lattice spacing should not be attributed to a gradual shift of the lattice constant with the incorporation of PFP molecules in the PEN crystal lattice as suggested in [Sal+08b]. Instead, it is likely that the absence of higher order reflections lead to some deviations in the results of the surface displacement error correction and hence, the obtained lattice constant is attributed to the PEN thin film phase as well. Together with the high crystallinity indicated by the Laue oscillations around the first order Bragg reflection it is concluded that the growth of the PEN crystallites is not disturbed by the presence of a comparably low number of PFP molecules.

The lattice distance of $d = (16.0 \pm 0.4)\,\text{Å}$ obtained for the PFP thin film is consistent with the literature data on the PFP thin film phase, characterized by a 15.7 Å out-of-plane lattice distance [Sal+08a][Kow+08]. This spacing is associated with the [100] crystal direction which approximately coincides with the long molecular axis similar to the PFP bulk crystal phase [Sak+04].

4.2.2. Electronic Transitions

The absorption studies were performed in transmission geometry and the obtained data was corrected for the transmission data of a plain glass slide yielding the neat transmission spectra of the molecular thin films. The absorption coefficient α has been extracted according to equation (3.16) and the transition spectrum has been calculated by $\bar{I} = \alpha/E$. Figures 4.3 a) to c) show the transition spectra of the low energy absorption spectra of the three samples and d) depicts a zoom-in of the lowest transition of all three samples on a semi-logarithmic scale. For all samples a decomposition of the signal using a linear combination of six Gaussians to fit the data has been performed. The individual peak functions (dashed lines) as well as the resulting cumulative fit (solid line) are presented in figure 4.3 a) to c) in addition to the data. The extracted central energies E_c as well as the FWHM Δ of the individual peaks marked in figure 4.3 with the respective label (ID), are listed in table 4.2. Annotations to the fitting procedure and value determination are marked by symbols at the respective label and are explained in the table caption.

For the PEN thin film, the absorption spectra associated with the $S_0 \rightarrow S_1$ transition comprises a low energy doublet (P_1 and P_2) and a series of peaks with decreasing intensity towards higher energies (V_1 - V_4) being fully consistent with literature [HHB80][Ost+05][Fal+06][Hin+07]. The P_1-P_2 doublet is commonly associated with a Davydov splitting originating by the two transitionally nonequivalent molecules in the PEN unit cell with an energetic splitting of approximately (110 ± 10) meV [HHB80]. The magnitude of the reported energetic splitting depends on various parameters. With increasing crystallinity Lee and Gan found an increase of the splitting from ≈ 110 meV to 135 meV [LG77]. An increase with decreasing temperature from ≈ 120 meV to 130 meV [Ost+05] and from 100 meV to 125 meV [Fal+06] has been observed, and a dependence of the herringbone angle in the various polymorphs has been reported [Mey+16]. The energetic splitting of 100 meV found for P_1 and P_2 as well as their peak energies correspond well to the reported literature values for the present PEN thin film phase [Fal+06][Hin+07][Mey+16]. It has been argued that the two lowest lying states cannot

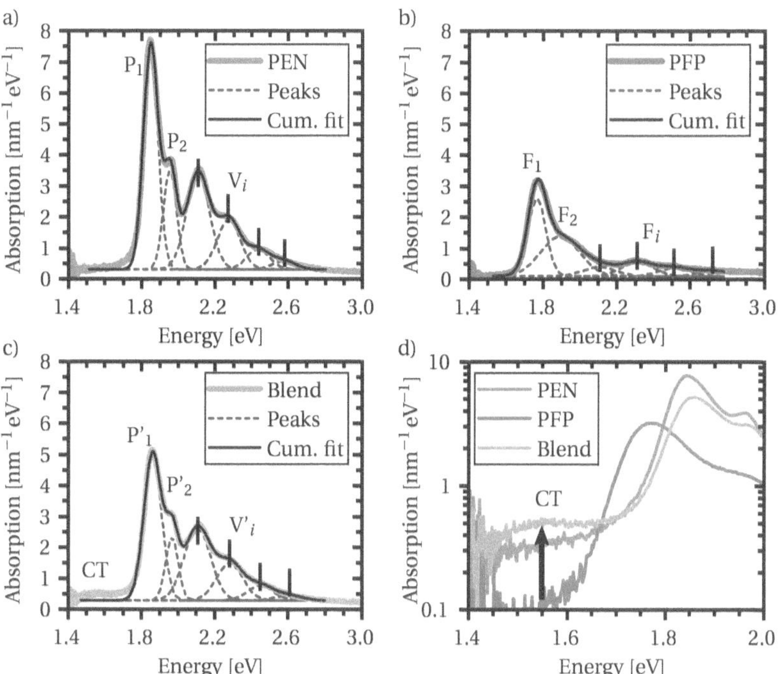

Figure 4.3.: Absorption presented as transition spectra for a) neat PEN , b) neat PFP, and c) blended PEN:PFP thin films including decomposition with Gaussians. d) Semilogarithmic plot of the low energy regime to highlight the CT transition in the blended film at 1.6 eV.

by attributed to a Frenkel exciton split in energy by a Davydov mechanism, but rather are the result of an additional low lying CT state [Gro+06]. However, in 2008 Dressel *et al.* showed by means of a Kramers-Kronig consitent interpretation of ellipsometry measurements on pentacene single crystals quite convincingly that the polarization dependence of the lowest lying excitation doublet is consistent with a Davydov splitting [Dre+08], although more recent calculations suggest a CT admixture of up to 50 % to the Frenkel exciton [Hes+15]. The need for a CT contribution to the Frenkel exciton to explain the observed values for the Davydov splitting in oliogacene crystals has been predicted before [Yam+11]. The peaks V_1 - V_4 are usually attributed to vibronic progressions [Gro+06][Hin+11][Hes+15], which are observed in crystalline aggregates if the inter molecular coupling is small compared to the vibrational coupling (*c.f.* figure 2.15). The energetic distance between V_1 to V_4 is (160 ± 10) meV, and thus beeing consistent with literature. The participating vibrational modes are commonly associated with C−C stretching and in-plane C−H bending modes [Gro+06]. This is consistent in terms of energies with the dominant in-plane C−H bending and *Kekulé* modes found by Raman spectroscopy [YKO07]. Assuming just one dominant vibrational mode, the peaks P_1 and P_2 comprise the 0-0 transition, while the successive peaks V_i are the related vibronic transitions 0-*i*. Then, the integrated intensities of the transitions normalized to the overall intensity directly resemble the respective Franck-Condon factors as the transition spectrum is a direct measure of the respective transition dipole moment. Together with the summed up intensities of P_1 and P_2 giving the 0-0 transition strength, the Franck-Condon factors are plotted in figure 4.4 a) (green filled circles). For a single dominant transition, the intensity distribution follows a Poisson distribution according to equation (2.38). Fitting this to the experimental data with the Huang-Rhys parameter S_{PEN} being the only free parameter reveals a good agreement (green open circles) and yields $S_{PEN} = 0.645 \pm 0.092$ being in good agreement within the error with the theoretical value $S_{PEN}^{DFT} = 0.574$ from density functional theory (DFT) calculations [Ste+16].

The PFP absorption spectra shown in figure 4.3 b) reveals a lower transition strength for the respective $S_0 \rightarrow S_1$ transition compared to neat PEN, being consistent with literature [Hin+07]. The ratio of the integrated intensities reveals that the PFP transition strength is only 46 % of PEN. Furthermore, the FWHM of the absorption peaks is consistently larger than that of the PEN film suggesting a higher energetic disorder. This is most likely caused by the overall lower crystallinity of the PFP film indicated by the XRD data. The energy of the lowest component F_1 [Hei+10] as well as the general shape of the spectra [Hin+07][BW11] are consistent with literature. The peak intensity ratio to the integrated intensity, as analyzed for PEN above, is plotted in 4.4 b). The inten-

PEN			PFP			Blend		
ID	E_c [eV]	Δ [eV]	ID	E_c [eV]	Δ [eV]	ID	E_c [eV]	Δ [eV]
P_1	1.85	0.09	F_1	1.78	0.11	P'_1	1.86	0.11
P_2	1.95	0.09	F_2	1.89	0.25	P'_2	1.97	0.08
V_1^*	2.11	0.15	$F_3^{*,\dagger}$	2.11	0.21	V'^*_1	2.11	0.17
V_2^*	2.27	0.15	F_4^*	2.31	0.21	V'^*_2	2.28 ± 0.01	0.17
V_3^*	2.44 ± 0.01	0.15	F_5^*	2.52 ± 0.01	0.21	V'^*_3	2.45 ± 0.02	0.17
V_4^*	2.58 ± 0.01	0.15	F_6^*	2.72 ± 0.01	0.21	V'^*_4	2.61 ± 0.03	0.17
						CT^+	1.57 ± 0.04	

Table 4.2.: Central transition energies E_c and FWHM Δ extracted from the decomposition of the transition spectra designated by their respective labels (ID) as in figure 4.3. Annotations to value determination as follows: * shared line width in fitting procedure, † fixed during fitting procedure, + estimated with reading precision from semi-logarithmic plot.

sity distribution is clearly not compatible with a Poisson distribution as for the two peaks in the transition spectrum F_2 and F_4, the intensity distribution shows two distinct maxima. Even if considering peak F_1 and F_2 as Davydov components of the same transition, the resulting intensity distribution would contain an additional maximum at peak F_4. This leads to the conclusion that a vibronic progression with just one dominant intramolecular mode as suggested by [Hin+07] is not sufficient to explain the experimental data, even though the energetic spacing is regular with (210 ± 10) meV and the energetic spacing fits a symmetric deformation mode of the two outer C rings with transition dipole moment oriented perpendicular to the molecular axis and being of sufficient intensity as suggested by Raman studies [Bre+12]. This is supported by the reported perpendicular polarization dependency of the transitions associated with F_1 to F_3 (group 1) and F_4 to F_6 (group 2) [BW11] which also contradicts the hypothesis that F_1 and F_2 are solely Davydov components as they have the same polarization dependency. An adequate explanation based on temperature and polarization dependent absorption measurements is given by Kolja Kolata [Kol14]: The transition F_1 contains three distinct transitions, two of them being Davydov components with an energetic spacing of about 25 meV, which is much smaller than the splitting observed in PEN due to the lack of CT admixture in the PFP Frenkel state. The third transition belongs to a resonant state delocalized along the b-axis of the crystal coinciding with one of the Davydov components. The peak group 1 is caused by a J-like coupling within the slip-stack aggregate along the b-axis leading an overall red shift. Hence, F_2 and F_3 can be considered vibronic progression, appearing damped by the high intermolecular coupling and the resulting enhancement of the 0-0 transition [Spa10], composed of two perpendicular polarizations related to the Davydov components. The peak group 2 results by an H-like coupling along the crystallographic c-axis reducing the overall transition strength and leading to a blue shift of the group of vibronic transitions. The resulting transition dipole moment along the c-direction leads to the distinct polarization.

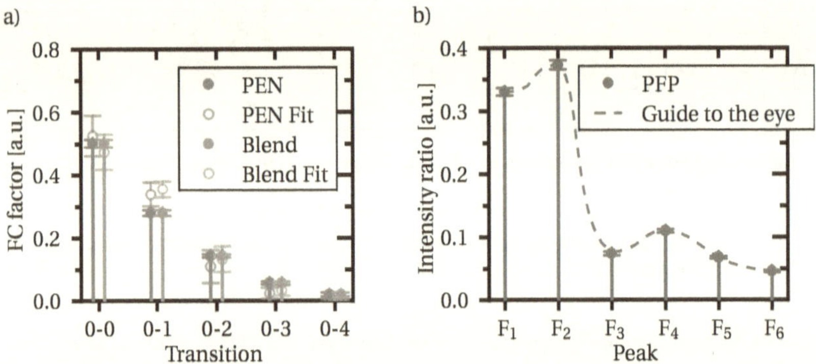

Figure 4.4.: a) Experimental (solid) as well as fitted (open) Franck-Condon factors for PEN (green) and blendend PEN:PFP films (orange). b) Intensity ratio of individual PFP transitions showing a clear deviation from a Poisson distribution.

The absorption of the blended PEN:PFP film is similar to that of neat PEN, although, the overall absorption is reduced to 77 % and the lowest peaks P'_1 and P'_2 are slightly blue shifted compared to PEN. This has also been observed for other low concentration admixtures of PFP to PEN [Bro+11]. The slightly higher FWHM compared to PEN, *i.e.* 10 meV to 20 meV, for all peaks could be related to a locally enhanced disorder in the environment of the PFP entities caused by the larger van der Waals radii of the fluorine ligands. Interpretation of the peak origins analogous to PEN leads to Franck-Condon factors beeing consistent with a vibronic progression originated by a single vibrational mode as depicted in figure 4.4 a) (orange). The free parameter Poisson fit reveals a Huang-Rhys parameter of $S_{\text{Blend}} = 0.749 \pm 0.079$. This Huang-Rhys parameter as well as the vibronic spacing of (160 ± 15) meV are consistent with the neat PEN film within the errors. This further supports the hypothesis of an absorption spectra mainly caused by PEN. For PEN-PFP dimers, a CT complex formation is observed which is most likely masked by the strong PEN absorption [Bro+11]. However, a small peak being absent in the neat PEN or PFP films can be found at ≈ 1.57 eV as indicated in figure 4.3 d). This peak has been associated with the absorption by a CT state [Bro+11][Kim+19] and can be confirmed even for small PFP fractions embedded in a crystalline PEN matrix.

4.3. Crystal Interface Morphology

The head-to-tail arrangement at the donor-acceptor interface can be realized by using a PEN (001) single crystal as a substrate for the subsequent PFP thin film growth. The crystal surface as well as the film growth have been characterized by complementary XRD and AFM studies to gain insights on the interface's molecular configuration.

4.3.1. Crystal Growth and Habit

PEN crystals have been grown by horizontal physical vapor deposition as described in 3.2.1 from twofold sublimation purified material. About 50 mg PEN were placed in an aluminium evaporation boat and sublimed at (285 ± 5) °C for 95 hours[1] in a 30 sccm N_2 inert gas flow. The temperature profile along the tube is depicted in figure 4.5 a) where the 0 cm position on the abscissa indicates the junction of the two heating coils (*c.f.* section 3.2.1). Crystals grew in a 2 cm wide recrystallization zone in the tube which corresponds to a temperature range of 110 °C to 140 °C. The grown crystals have been slowly cooled to room temperature within 12 hours to minimize thermal stress.

The obtained crystals preferentially exhibit an elongated plate like growth with several mm in length and varying habit. Alternatively, a dendritic growth can occur as well. Representative crystals are shown in figure 4.5 b) to d) as photographs (left) and optical microscope images at ten fold magnification (right). Obviously, irregularities in the crystal habit tend to be accompanied by an irregular surface structure. For example, the small needle-like crystal in b) shows a smooth surface with only few defects parallel to each other and perpendicular to the crystal edges. In contrast, the crystal surfaces presented in c) and d) exhibit, in addition to the seemingly rougher surface, step edges which enclose growth domains rotated with respect to each other, dislocations presenting as grooves in the surface, and twin domains which can be identified by their reflection using a polarizer to analyze the reflected light as done for the crystal shown in c). There, the polarized light reveals some domains (blueish) with a distinct orientation from the main crystal (reddish) on the scale of several 100 µm. As will be revealed by means of AFM height data in the next section, these macroscopic displacements do hardly translate to a similar defect density on the microscopic scale as could be expected.

Interestingly, all crystal types presented above can appear in the very same growth cycle suggesting that the local growth environment in the tube has a strong influence

---

[1]The uncertainty of the sublimation temperature is not related to fluctuations in the sublimation temperature but to an uncertainty in the measurement performed outside the inner tube to avoid creation of an artificial condensation point.

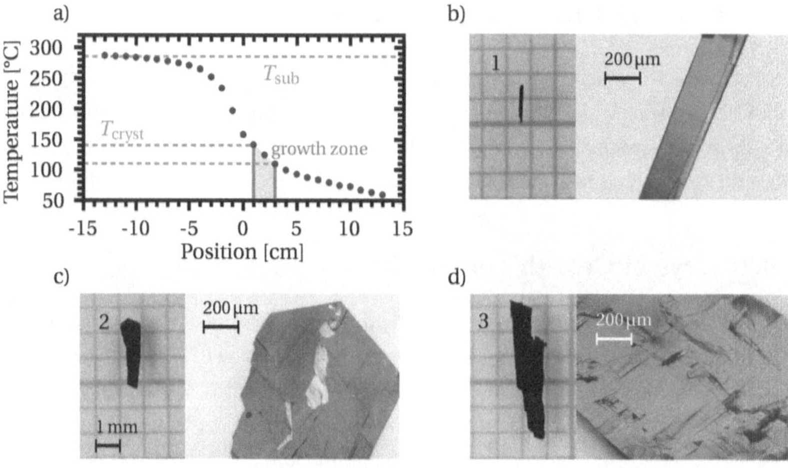

Figure 4.5.: a) Temperature profile along the growth tube during PEN crystal growth. Sublimation temperature T_{sub} as well as recrystallization temperature range T_{cryst} marked by horizontal lines. b)-d) Photographs (left) and optical microscope images, the latter at ten fold magnification, (right) of exemplary PEN crystals. Enumeration intended for later identification. b) Microscope image reprinted with permission from [Ham+20]. Copyright 2020 American Chemical Society.

on the growth of the individual crystals and can hardly be controlled by the global parameters such as temperature gradient, growth time and transport gas flow. As discussed in section 2.2.3, the emergence of a particular crystal facet depends on its growth speed and the respective kinetics of all possible crystal facets which, in turn, depend, *inter alia* on the crystallization temperature, the binding energy on the respective surface in relation to the binding energy inside the crystal, and the degree of supersaturation of the vapor phase. In addition to that, local turbulences in the buoyancy driven convection as well as local temperature fluctuations are likely to lead to growth scenarios far from thermal equilibrium such as dendritic growth. For example, an overall larger crystal growing over the same time span is likely to experience a locally enhanced condensation rate and hence, the deposition of more material in the same time. This can lead to a macroscopic surface roughening as the material diffusion to step edges and the subsequent agglomeration at those becomes the dominant process over the Kossel facet growth.

4.3.2. Structure and Surface Morphology

The three crystals shown in figure 4.5 b) to d) have been analyzed via XRD measurements. The corresponding diffractograms are shown in figure 4.6 a) for the crystals labeled 1, 2, and 3. All three crystals show four main (00l) Bragg reflections at $0.444\,\text{Å}^{-1}$, $0.889\,\text{Å}^{-1}$, $1.335\,\text{Å}^{-1}$, and $1.780\,\text{Å}^{-1}$ accompanied by low intensity reflections at $0.437\,\text{Å}^{-1}$, $0.875\,\text{Å}^{-1}$, $1.313\,\text{Å}^{-1}$, and $1.750\,\text{Å}^{-1}$ presenting as subsidiary peaks to the main (00l) reflections in figure 4.6 a). For crystal 1 the subsidiary reflections were only detectable next to the (001), (002), and (003) reflections. Figure 4.6 b) shows a zoom-in of the (003) reflection of crystal 1 together with the subsidiary peak at lower momentum transfer values with respect to the main peak. PEN is known for its polymorphism (*c.f* section 3.1.1) and hence, the reflection doublet at each (00l) Bragg peak likely originates by two distinct polymorphs. For a clear distinction of the two polymorphs associated with the peak doublet, the one associated with the reflection at higher momentum transfer will be labeled P1 while the other one will be labeled P2. To determine the characteristic out-of-plane lattice distance d_{001}, the extrapolation method for eliminating surface displacement errors ($\propto \cot\theta\cos\theta$) as well as instrumental and statistical errors ($\propto \cot\theta$) has been applied [KP01]. The extrapolation plots referenced to a $\cot\theta\cos\theta$-abscissa for all crystals and both polymorphs P1 and P2 are shown in figure 4.6 c). The analysis in terms of the $\cot\theta$-abscissa yields the same results for the given precision and is not shown for this reason.

	P1 d_{001}	P2 d_{001}	
Crystal 1	14.11 Å	14.34 Å	
Crystal 2	14.12 Å	14.36 Å	
Crystal 3	(14.15 ± 0.01) Å	14.38 Å	
Theoretical	14.13 Å [Sie+07]	A: 14.24 Å	B: 14.37 Å

Table 4.3.: Experimental (001) lattice constants for both pentacene polymorphs P1 and P2. Theoretical values are determined with the PowderCell software from published crystal structures. For P2, value A was obtained by calculating the room temperature unit cell lengths from data published in [Sie+07] and value B was determined from [Mat+01].

Table 4.3 lists the deduced lattice constants d_{001} for all three samples (crystal 1 to 3) and both polymorphs (P1 and P2), while the last row contains theoretical values calculated with the program PowderCell [KN96] from lattice parameters reported in literature. For P1, all samples comply with the bulk crystal phase associated with the 14.1 Å lattice distance along the crystallographic [001] direction and the theoretical value cal-

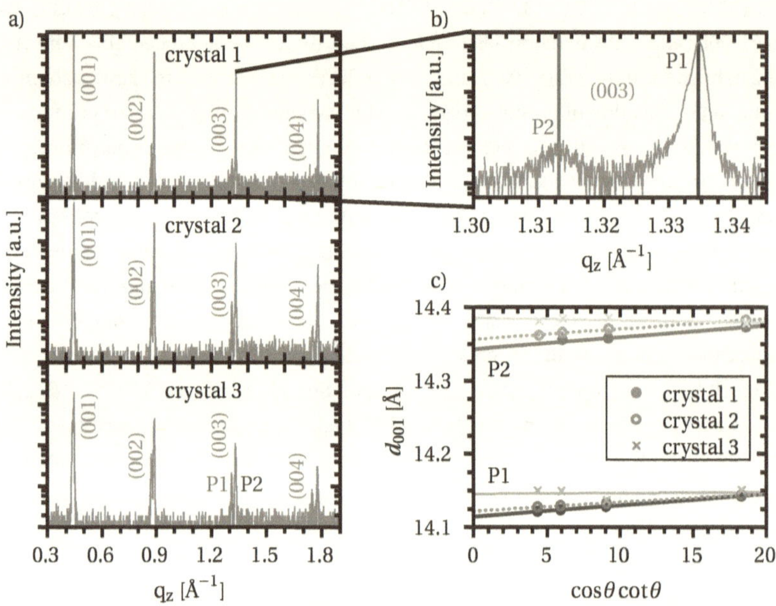

Figure 4.6.: XRD analysis of the three PEN crystals shown in figures 4.5 b) to c). a) XRD pattern for all three crystals comprising main Bragg reflections accompanied by low intensity subsidiary peaks forming a peak doublet at each (00l) reflection. Each peak is associated with a distinct polymorph, labeled P1 and P2, for higher and lower momentum transfer, respectively. For the (003) diffraction peak of crystal 3, the doublet peaks are labeled by the respective polymorph. b) Zoom-in on the (003) peak doublet of crystal 1 comprising reflections of the polymorphs P2 (lower q_z) and P1 (higher q_z). c) Determination of the respective lattice constant by extrapolation to 0 on an $\cos\theta\cot\theta$-abscissa.

culated from [Sie+07] is in good agreement with this data. Asking for the underlying crystal structure of the P2 polymorph, two hypotheses are proposed:

A - Residual high temperature phase: Siegrist *et al.* have shown that a coexistence of the bulk crystal phase and its high temperature analogue is possible even at room temperature and published the crystal structure for this high temperature phase as well as its thermal expansion coefficients [Sie+07]. From the published crystal structure the unit cell lengths of the high temperature phase at room temperature can be estimated by employing the thermal expansion coefficients while simultaneously keeping the unit cell angles fixed. This procedure yields a (001) lattice spacing of 14.24 Å.

B - Coexisting thin film polymorph: A thin film phase with a lattice distance of 14.37 Å has been reported, too [Mat+01]. It is stable over a wide temperature range [Mat+03b] but, to the author's best knowledge, has not been reported for single crystals so far.

Hence, Both models are suitable to explain the measured lattice distance for the second polymorph. The lattice distance of 14.24 Å obtained by hypothesis A is about 0.1 Å below the experimental value. However, for the calculation the unit cell angles were kept at a constant value. This assumption can only be justified as an approximation, as in a molecular crystal the van der Waals interaction between the molecules most likely changes the unit cell angles if the inter molecular distances change with temperature. A change in the angles α and β by 1° varies the lattice distance by about 0.1 Å, leading to an agreement with the experimental values. The transition temperature from the bulk phase (P1) to the Siegrist high temperature phase has been determined to 190 °C, however, the latter is metastable at room temperature [Sie+07]. Thus, it seems plausible that in the observed temperature range of the crystal growth the high temperature phase can form stable configuration and, thereafter, the slow cooling can result in co-exisitng metastable crystallites attached to the crystal, *e.g.* on the crystal surface. The lattice distance of 14.37 Å associated with a thin film phase is in good agreement with the observed (001) value. A possible growth mechanism inducing the growth of the thin film phase could be related to the slow cooling procedure. At the beginning of the cooling process, material is still sublimed, even though, the temperature at the position of the starting material changes which leads to a change of global growth parameters in the recrystallization zone, *i.e.* the degree of supersaturation and the temperature gradient. The formation of a thin film phase cover layer on top of the existing bulk crystal is unlikely. This would require that the reduction of the Gibbs potential during the thin film phase condensation compensates the interface energy at the phase boundary, which, in the case of ongoing bulk phase growth would be non-existent. Hence, the thin film growth should be energetically less favorable due to the existence of a phase boundary. However, nucleation of a different phase at already existing surface defects acting as condensation seeds could lift this energetic constraint. At this point, a conclusive explanation cannot be deduced from the present data. As in both polymorphs the [001] crystal direction coincides with a standing molecular configuration at the crystal surface, the use of PEN (001) single crystals as templates for a standing stack aggregate formation is not impaired by the occurring polymorphism.

To confirm the presumed upright orientation of the PEN molecules at the (001) crystal surface, complementary AFM studies on the crystals 1, 2, and 3 have been performed. The AFM height data of representative $5\,\mu m \times 5\,\mu m$ sections on each crystals' surface are depicted in figures 4.7 a), e), and f) for crystals 1, 2, and 3, respectively. It is clearly visible that all crystals show large smooth terraces separated by mono molecular steps. The vicinal surface of crystal 1 shows a regular growth pattern with a distinct step flow growth direction, while the crystal surfaces of crystals 2 and 3 exhibit a high number of kinks in the step edges. The root-mean-square surface roughness determined on a representative terrace amounts to $0.61\,\text{Å}$, $1.25\,\text{Å}$ and $7.57\,\text{Å}$ for crystals 1, 2, and 3, respectively, showing the high surface quality. Even in case of crystals 2 and 3, for which macroscopic surface inhomogenities have been observed in the optical microscope images (*c.f.* figure 4.5), the AFM data indicates large (001) planes interrupted only by mono molecular steps. For crystals 2 and 3 in figures 4.7 e) and f) the self-affinity of the macroscopic inhomogenities on the crystal surfaces with the saw shaped terraces and the small kinks in the step edges suggests the local step flow growth to be diffusion limited [WS83].

Figure 4.7 b) displays a zoom-in of the vicinal surface of crystal 1, revealing small kinks in the steps. Measuring the angle enclosed by the extension of the step edge and the kink, as indicated by the white lines in figure 4.7 b), at a total of nine similar locations yield an average kink angle of $(81.9 \pm 3.3)°$ (arithmetic mean $\pm$ standard deviation). Comparing this value to reported unit cell angles reveals an agreement with the γ angle of the bulk crystal phase of $\gamma = 84.7°$ [Mat+01][Sie+07], associated with polymorph P1 as well as with the γ angle of the thin film phase of $\gamma = 80.9°$ [Mat+01][Mat+03b], associated with the previously hypothesized polymorph P2. Furthermore, the γ angle of the Siegrist high temperature phase, $\gamma = 85.9°$ [Sie+07], *i.e.* of another suggested crystal structure of polymorph P2, is close to the observed value. Although no clear distinction between the three polymorphs found by XRD can be made at this point, assuming the deduced angles are a manifestation of the unit cell geometry, crystallographic directions can be assigned to the step edges. Assuming the mea-

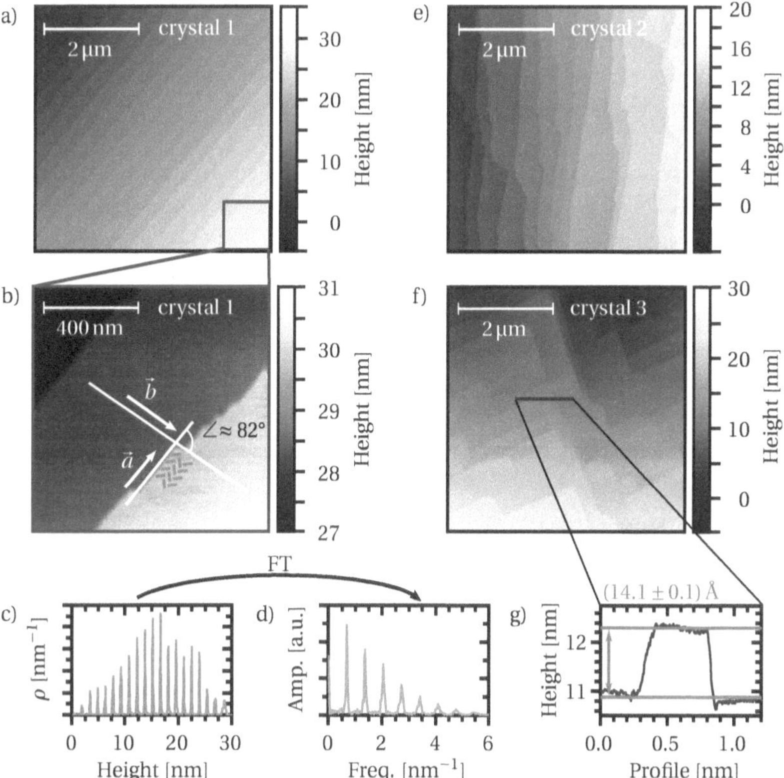

Figure 4.7.: AFM scans of PEN (001) crystal surfaces. a), e) and f) show 5 μm × 5 μm scan sections of the surfaces of crystal 1, 2, and 3, respectively. b) A 1 μm × 1 μm zoom-in of a) is depicted and the angle of an exemplary kink at the step edge is determined. White arrows indicate the respective crystallographic directions and the green rectangular dots indicate the expected molecular packing. c) Height distribution at the surface of crystal 1 and d) amplitude of the related Fourier transform. g) Height profile across a mono molecular step on the surface of crystal 3 with an estimated step height of (14.1 ± 0.1) Å.

sured γ angle corresponds to that of the unit cell, the direction of the step edge agrees with the crystallographic $\vec{a}$ axis, whereas the kink edge, respectively, corresponds to the crystallographic $\vec{b}$ direction. The green rectangular dots in 4.7 b) schematically (not to scale!) display the molecular pattern from a top view. As the (ac) surface, related to the respective edge, extended much larger than the (bc) surface, this indicates a faster growth along $\vec{a}$ direction, which correlates with an in-plane molecular slip stack pattern as highlighted in 4.7 b). In contrast, the growth speed along the $\vec{b}$ direction corresponding to an in-plane herringbone pattern is much slower.

The representative height distribution of crystal 1 (figure 4.7 c)) reveals a discrete distribution with equidistant peaks, confirmed by the frequency spectrum obtained by the corresponding Fourier transform shown in figure 4.7 d). Fitting a Gaussian distribution to the first frequency peak reveals an average step height of $(14.6 \pm 1.0)\,\text{Å}$ confirming the mono molecular step height of the vicinal surface. Still, a distinction between the three polymorphs discussed above is not possible based on the data due to the width of the frequency distribution. An identical analysis of the other data sets reveals similar results, even though the discrete steps are not as pronounced as in the height distribution of crystal 1. Based on profile lines, exemplary shown for crystal 3 in figure 4.7 g) and fitted by a step edge function a step height of $(14.1 \pm 0.1)\,\text{Å}$ can be estimated, in good agreement with the PEN bulk crystal phase.

The pentacene molecule is known to be prone to oxidation, with e.g. two oxygen atoms binding to the inner carbon ring effectively breaking the π-system and forming *6,13-pentacenequinone* $C_{22}H_{12}O_2$ [De +09]. A pronounced oxidation of the crystal surface would question the use as substrates for the subsequent PFP growth. The distinct step height of $17.79\,\text{Å}$ for 6,13-pentacenequinone has been used as an indicator to determine the presence of oxidized species on similar single crystal surfaces by means of AFM studies before. There, about 17 % of the probed crystal surface was oxidized [Jur+07]. Studies on PEN thin films suggest that exposure to air or to molecular oxygen only leads to irreversible oxidation under simultaneous exposure to UV-light, which is attributed to the generation of highly reactive oxygen species, *e.g.* singlet oxygen [Vol+05] [Vol+06]. This rapid oxidation was accompanied by a dissolving of the PEN thin film and no evidence of the formation of stable quinone species was found [Vol+06]. As the exposure to ambient light has been kept to a minimum during sample handling and neither an indication of the fingerprint 6,13-pentacenequinone step height or any indication of surface reactions have been observed in the presented AFM data, it can be concluded that surface oxidation is not a pronounced effect on the present crystal surfaces.

4.3.3. Donor-Acceptor Crystal Interfaces

With the purpose of a structural analysis of the PFP film growth on top of a PEN (001) single crystal surface a high quality PEN single crystal, similar to the previously presented crystal 1, was chosen. The following XRD analysis has in part been published in [Ham+20]. The crystal was mounted stress free on a precleaned glass slide by means of Kapton® tape for the up side down vacuum deposition of PFP on top. Prior to PFP deposition, the out-of-plane Bragg reflection of the crystal was measured and the XRD pattern is shown in figure 4.8 a). The high crystal quality is confirmed by the intense Bragg reflections up to the eighth order and the sharp rocking peak with a FWHM of 0.063° at the second order Bragg reflection (*c.f.* figure 4.8 c)). The enhanced background between $0.3\,\text{Å}^{-1}$ and $2.5\,\text{Å}^{-1}$ is attributed to the diffuse scattering by the Kapton® tape. An extrapolation of the deduced lattice constant of each reflection peak to correct for a possible sample displacement [KP01] yields a corrected (001) lattice spacing of $d_{001} = 14.13\,\text{Å}$, matching the expected bulk crystal phase. Again, a low intensity satellite peak has been observed for the first two Bragg reflections in accordance to the PEN polymorph P2.

PFP was deposited via high vacuum sublimation onto the PEN (001) crystal surface with a nominal thickness of 20 nm at an average deposition rate of $0.17\,\text{Å}\,\text{s}^{-1}$. Repeating the same XRD measurement as for the uncovered PEN (001) surface reveals a weak intensity peak at $q_z = 0.395\,\text{Å}^{-1}$ (*c.f.* figure 4.8 b) corresponding to a lattice spacing of $d = (15.9 \pm 0.2)\,\text{Å}$. This is in good agreement with the (100) lattice distance of the PFP thin film phase [Sal+08a] and corroborates the intended molecular head-to-tail arrangement at the PFP/PEN crystal interface.

As a XRD analysis only yields integral information on the sample's structural data AFM studies at a sub monolayer, a 10 nm, and a 20 nm PFP film thickness, were performed. Figures 4.9 a) and b) depict the AFM height and phase data of a monolayer PFP island on a PEN terrace. From the line profile in the height data shown in 4.9 c), the PEN step edge can be clearly identified by the height difference of 14.0 Å, which corresponds to the PEN (001) lattice spacing. The island on the right side of the AFM image is characterized by a height difference of 15.6 Å (*c.f.* figure 4.9 c), in agreement with (100) PFP lattice distance. The phase data presented in 4.9 b) shows a constant value for the attributed PEN crystal surface, only changing when crossing a step edge at the monolayer island. Analyzing a profile line at the same position as for the height data illustrates the clear distinction between the two areas. As a change in the tip's oscillation phase indicates a change in energy dissipation to the surface (*c.f.* section 3.3.1), it can be concluded that the interaction between the tip and the material comprising the island differs from the interaction with the uncovered PEN crystal surface. Together

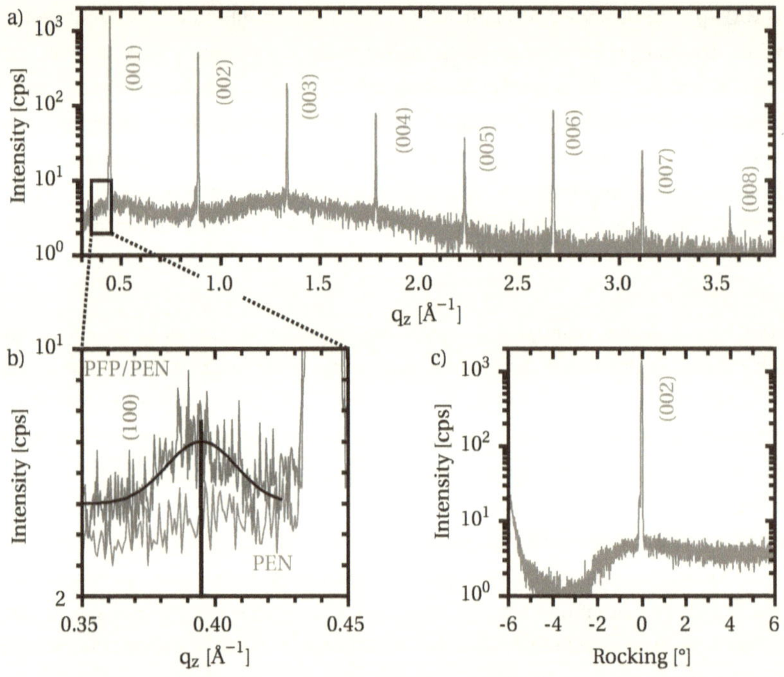

Figure 4.8.: a) XRD pattern measured on an uncovered PEN single crystal. b) Zoom into the XRD pattern prior and after 20 nm PFP deposition on the same crystal revealing a (100) PFP thin film reflection. c) Rocking curve at the (002) PEN Bragg peak. Adapted with permission from [Ham+20]. Copyright 2020 American Chemical Society.

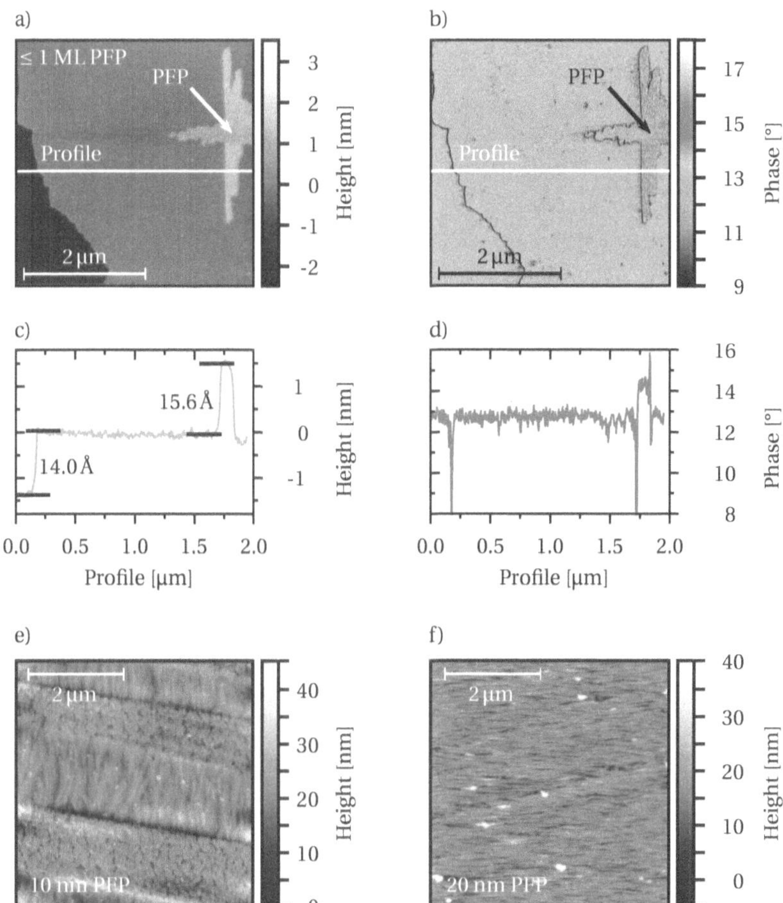

Figure 4.9.: AFM studies of PFP film growth on PEN (001) single crystal surface. a) and c) show the height and phase image of a monolayer (ML) PFP island (labeled by PFP) on a PEN (001) single crystal surface. A profile has been extracted at the same position of each data set. c) and d) show the height and phase profiles along the lines marked white in the scans of a) and c). The step edge in the lower left of the AFM height data in a) can be assigned to the PEN (001) crystal surface, as no change in phase contrast across the edge is apparent and the mono molecular step height of 14.0 Å can be assigned to the PEN bulk phase. For the island in the right of the AFM height data a clear phase contrast as well as a step height of 15.6 Å identifies the present molecules as PFP. The distinct mono molecular out-of-plane step heights of both materials confirm the molecular head-to-tail orientation at the interface. e) and f) show AFM surface data of a 10 nm and 20 nm thick PFP film on a PEN (001) crystal surface. a)-d), f) adapted with permission from [Ham+20] and the corresponding supporting information. Copyright 2020 American Chemical Society.

with the fact that island's height matches the (100) PFP lattice constant, the island can unambiguously assigned to a PFP layer growing on top of the PEN (001) crystal surface. The step heights of the PEN and PFP terrace indicate that already directly at the interface the molecules stack in a head-to-tail orientation. This is an important observation as for other D-A systems, *e.g.* copper-phthalocyanine and its perfluorinated counterpart, a reorientation at the interface, leading to a π-stacking and a resulting cofacial arrangement, has been observed [Opi+16]. From the combined XRD, AFM height, and AFM phase data such a reorientation can be excluded.

The AFM data of PFP films of 10 nm and 20 nm thickness grown on top of the PEN (001) crystal surface of and shown in figures 4.9 e) and f), respectively, indicate that a full coverage of the surface could be achieved. The thin films are composed of large domains comprising highly orientated needle like crystallites up to 1 µm in length. The evaluation of the corresponding profile lines on the 20 nm film yields a mono molecular step height in agreement with the (100) PFP lattice spacing, further corroborating an upright orientation of the PFP molecules as expected from the above presented data. The step line analysis has been published in [Ham+20] as supporting information and the relevant excerpt can be found in appendix C.

4.4. Optical Studies of Donor/Acceptor Crystal Interface

Optical investigation of CT states and complexes constitutes a well established approach to characterize their energetics and dynamics. For this purpose, the single crystal PFP/PEN interface has been examined be means of steady state photoluminescence (PL) and differential reflectance spectroscopy (DRS) to characterize the emission as well as absorption behavior of a potential CT complex forming in the head-to-tail geometry at the common interface.

4.4.1. Temperature Dependent Photoluminescence

The PEN single crystal and the silicon diode for temperature measurement and control were fixed with silver paint on a copper plate to ensure good thermal coupling to the cryostate's cold finger. Thermal contact between cold finger and copper substrate was provided by a thin layer of silver paint. The sample was cooled down to 5 K and than slowly heated up. PL spectra were recorded at fixed temperatures using a 300 lines mm^{-1} diffraction grid at central wavelengths of 625 nm, 750 nm, 875 nm, and 950 nm with an acquisition time of 30 s, respectively. Prior to the PL measurement the sample was held at the respective temperature for at least 15 min to ensure thermal equilibrium. An excitation wavelength of 532 nm was chosen. The excitation light was circular polarized by a $\lambda/4$ wave plate to exclude polarization effects during excitation by the pronounced Davydov splitting of PEN. A 550 nm long pass filter was added to the signal pathway to block residual laser light. The resulting transition spectra $\bar{I}(E)$ normalized to the integrated intensity of the neat PEN crystal for temperatures from 5 K to 295 K are shown as green lines in figure 4.10. After the measurement the sample was carefully removed from the cold finger and a 20 nm thick PFP thin film was deposited by physical vapor deposition at an average rate of $1 \,\text{Å} \text{s}^{-1}$. The PL measurement procedure was repeated with the same parameters. The results are depicted as the blue curve in 4.10 for comparison. Although PEN is prone to a fast exothermic singlet fission [Tho+09][Wil+11][Zim+11], the fluorescence is not fully quenched. However, the overall intensity is about 100 times lower compared to alike systems such as tetracene or rubrene measured under similar conditions.

The PFP/PEN spectra is modulated by a superimposed low energy oscillation below 1.6 eV. This is the result of a *Fabry-Pérot interference* where the two parallel crystal faces act as a resonator for the emitted light [Nie05]. The energetic distance between two adjacent transmission maxima ΔE is given by the *free-spectral range* which is expressed in terms of energy as

$$\Delta E = \frac{ch}{2nd} \tag{4.1}$$

with n beeing the refractive index, d the crystal thickness, and c and h are speed of light in vacuum and the Planck constant, respectively [Dem11]. For 5 K the energetic spacing between 1.5 eV to 1.7 eV is determined to $\Delta E \approx 6 \,\text{meV}$ and with a nearly constant refraction index of $n = 1.87$ in this spectral range [Dre+08] a crystal thickness of about 10 µm is obtained. This is in the range of reported crystal thicknesses for PEN crystals obtained by similar methods [Dre+08][Gro+06]. For the PL signal of the neat crystal an oscillation is only visible for 20 K with larger period of $\Delta E \approx 20 \,\text{meV}$ corresponding to a crystal thickness of *ca.* 3 µm. It has been observed that PEN crystals do not necessarily have a homogeneous thickness across their lateral extend and the signals of the respective measurements were not recorded at the exact same spot on the crystal surface. An additional smoothing of the sample's surface by the PFP cover layer could support the occurrence of the interference pattern as PFP has a similar out of plane refraction index $(n_{PFP} \approx 1.4)$ [Hin+07] in the respective energy range.

The key result of the PL comparison of the neat crystal and that covered by the PFP top layer is the absence of any detectable difference in the transition spectra over the whole temperature range. This is a noteworthy difference with respect to the optical data achieved on mixed films [Ang+12][Rin+17][Bro+17]. There, most of the PEN molecules' photoluminescence signal is quenched at low temperatures and a new transition at 1.4 eV is observed. There are, however, subtle differences in the relative intensities of the individual transitions between the two measurements. These can be explained by the nature of the respective transitions and, hence, are likely to be an inherent feature of crystalline PEN rather than a result of any interaction between PEN and PFP at the interface. The individual transitions will be discussed below to corroborate this hypothesis.

Temperature dependent studies of the luminescence behavior of crystalline PEN are rare in literature, and most PL analyses refer to two main publications [Aok+01] and [He+05]. The temperature dependent PEN spectra in figure 4.10 show four main transition peaks indicated by the dotted lines marking their energetic positions at 5 K and labeled as E, T, DT_1 and DT_2 from high to low energy. At low temperatures, E and T show an asymmetric spectral line width with a pronounced low energy flank which seems to vanish with rising temperature. Additionally, a slight red shift can be observed for E. All peaks exhibit a pronounced temperature dependence in their relative and overall intensity. As explained later in detail, some contributions might be the result of lattice deformations and the excimeric nature of the respective transition prevents a simple decomposition by using Gaussian functions. For this reason, a simple relation between the absolute intensity and the temperature can not be derived. However, it is possible to approximate this relation under the assumption that the temperature dependency of a transition is independent of its emission energy. This assumption is justified, if the underlying process is an external quenching process, *e.g.* non-radiative depopulation by exciton-phonon scattering, or a thermally activated redistribution of the state population and hence, does not directly influence the respective transition dipole moment, as it is usually the case in molecular solids. This means that any integral over the transition spectrum with constant integration boundaries is directly proportional to the temperature dependent transition rate

$$\overline{I}_{\mathrm{int}}(T) = \int_{E_1}^{E_2} \overline{I}(E; T)\mathrm{d}E \propto A(T)\left|\vec{M}^2\right| \tag{4.2}$$

where $A(T)$ is a temperature dependent function and $\vec{M}$ is the transition dipole moment of the respective transition. Choosing the integral boundaries carefully to maximize the signal to noise ratio and simultaneously exclude the interference with adjacent transitions as possible resulted in an integration width of 30 meV. The following integration ranges where chosen: 1.757 eV to 1.787 eV (E), 1.637 eV to 1.667 eV (T), 1.475 eV to 1.505 eV (DT_1) and 1.415 eV to 1.445 eV (DT_2). The results are shown in figure 4.11 for all four transitions as function of the respective thermal energy in terms of

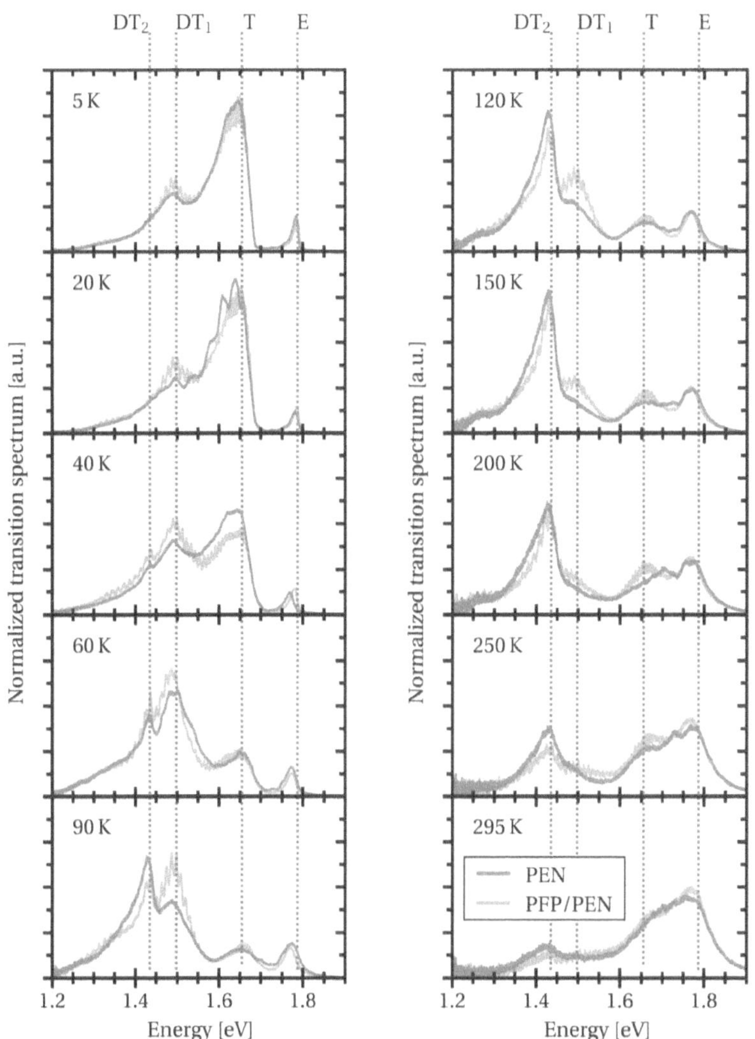

Figure 4.10.: Temperature dependent photoluminescence transition spectra for a PEN (001) crystal surface and a PFP/PEN (001) crystal surface. Each transition spectrum has been normalized to its integrated intensity for the sake of comparability. Main transitions labeled E, T, DT₁, and DT₂ are indicated by horizontal lines and discussed in the text.

$1/k_{\mathrm{B}}T$. A Poisson noise of $\sqrt{\overline{I}_{\mathrm{int}}}$ and an 1 K uncertainty in the temperature measurement were assumed. The three transitions at higher energies, *i.e.* E, T, and DT_1, could be modeled by the assumption of a thermally activated non-radiative transition in combination with a constant radiative transition rate A leading to

$$\overline{I}_{\mathrm{int}} = A - B\exp\left(-\frac{E_{\mathrm{a}}}{k_{\mathrm{B}}T}\right) \tag{4.3}$$

as a fitting equation. The rate model used to describe the temperature dependence of DT_2 is discussed in detail below. The extracted activation energies E_{a}, the energy position at 5 K, and the ascribed excitonic processes are listed in table 4.4.

Label	Position at 5 K	Activation energies	Attribution	Reference
E	1.785 eV	$E_{\mathrm{a}} = (12 \pm 6)\,\mathrm{meV}$	free exciton	[Aok+01][He+05]
T	1.65 eV	$E_{\mathrm{a}} = (2 \pm 1)\,\mathrm{meV}$	self trapped	[Aok+01][He+05]
DT_1	1.49 eV	$E_{\mathrm{a}} = (5 \pm 2)\,\mathrm{meV}$	deep self trapped H-defect impurity	[He+05] [Aok+01]
DT_2	1.43 eV	$E_{\mathrm{a}} = (17 \pm 1)\,\mathrm{meV}$ $E_{\mathrm{b}} = (9 \pm 2)\,\mathrm{meV}$	trap state at structural defect	not reported in literature

Table 4.4.: Four main transitions observed in the PEN single crystal PL together with their respective main energy position at 5 K, thermal activation energies and attributed excitonic processes.

The PL peak E is associated with the radiative decay of the free singlet exciton [Aok+01][He+05] and has been observed in thin films samples [Ang+12] on a 80 ps timescale [Bro+17] as well. Even though less pronounced, it can also be detected at low intensities in electroluminescence from PEN nano crystallites [Kab+10]. With rising temperature, the main transition energy experiences a slight red shift of 15 meV to 1.77 eV together with adapting the shape of a Gaussian emission profile. This can be explained by small static disorder. The transport of excitons is usually temperature activated [Top+14] and can be described by a Marcus-like hopping transport [Ste+14]. At cryogenic temperatures, excitons are mostly immobile while the hopping to adjacent sites is, if at all, preferably happening to low energy sites of the disorder distribution leading, together with the presence of only low energy vibrational progressions, to an asymmetric line shape as observed in the spectra of figure 4.10. With rising temperature, the barycenter of the spectral distribution shifts to lower energies, as the transport to the low energy sites is enhanced. For even higher temperatures the population distribution becomes continuously evened, leading to a symmetric emission profile. Figure 4.11 (left upper panel) shows a decrease in the overall emission intensity with temperature, and the fit by equation (4.3) resembles the course of the data well within the confidence bands. The extracted activation energy of $E_a = (12 \pm 6)$ meV is in agreement with literature values of (9.97 ± 1.42) meV [He+05], substantiating the applicability of the chosen analysis method. As singlet fission is an exothermic process in PEN [Zim+11][Yos+14], and hence can be excluded as a thermally activated decay channel of the singlet exciton. This leaves a phonon induced non-radiative transition as a possible explanation for the observed behavior with temperature. Indeed, the activation energy matches the frequency of a bulk phase optical phonon with a frequency of $92\,\mathrm{cm}^{-1}$ (≈ 11.4 meV) [Bri+02], further corroborating the phonon hypothesis.

The adjacent emission peak at 1.65 eV, labeled T, has been previously assigned to a self trapped exciton [Aok+01][He+05]. It has also been observed in high quality thin films [Ang+12] and is characterized by a 3 ns decay time constant [Rin+17]. In electroluminescence this transition is the dominant radiative decay channel [Par+02][SMS09][Kab+10]. Commonly, the formation of a self trapped exciton is attributed to a local lattice deformation due to polarization effects, but a lowering of the energy of a Frenkel exciton is unlikely due to its already high degree of localization and its charge neutrality [PS99]. For PEN however, the excited singlet state is associated with a strong admixture of a CT state [Hes+15][Yos+14] enabling a lattice polarization. This is corroborated by the dominance of this emission band in electroluminescence while the photoluminescence at room temperature is governed by the emission from the free excitonic state. The over all intensity of the self trapped state T is sharply decreasing with temperature with an activation energy determined to (2 ± 1) meV (*c.f.* figure 4.11 left lower panel). The fast depopulation can be explained by assuming that the exciton is highly sensitive to changes in its local surroundings caused by the nature of the self trapping process. Thus, dynamic disturbances due to lattice phonons are likely to influence this emissive state. Studies on the charge carrier phonon coupling predict a low energy optical phonon mode at around 3 meV to be strongly coupled to free charge carriers [Mas+04]. It can be speculated that this low energy optical phonon could also be responsible for the fast depopulation of the CT state induced self trapped exciton initiated by CT formation.

The origin of the transition at 1.49 eV, labeled as DT_1, is currently still under discussion. He and coworkers suggested an emission related to a 6-hydropentacene $C_{22}H_{15}$ impurity where a center ring carbon atom forms a sp^3 hybridization with a second hydrogen atom forming a CH_2 molecular defect which slightly disturbes the π-system [He+05]. The authors argue that in thin film transistors, hole trap states have been observed at 0.35 eV above the transport level, closely matching the observed energetic shift from the free exciton luminescence to the DT_1 signal [Lan+04]. This rise

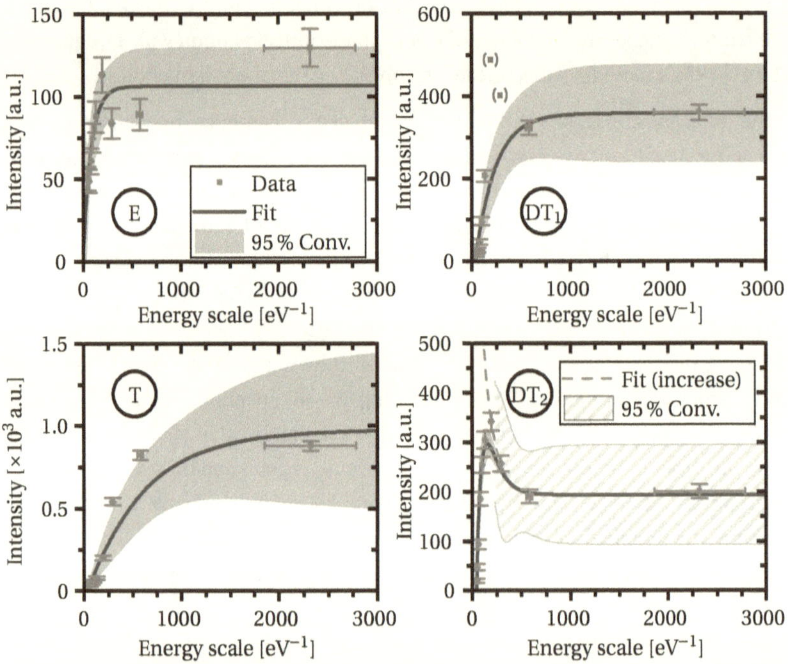

Figure 4.11.: Temperature dependent intensities of the four main transitions labeled E, T, DT_1 and DT_2 as function of thermal energy scale in terms of $1/k_B T$ including fit (green) and corresponding 95 % confidence bands (grey). The fit of the intensity rise for the DT_2 emission at low temperatures is given by the grey dashed line together with its confidence band (grey dashed area). Discarded data points of the DT_1 intensity attributed to interference with the DT_2 emission band are marked in red and set in brackets.

in HOMO energy matches the theoretical predictions for the neutral $C_{22}H_{15}$ or the cationic $C_{22}H_{15}^+$ species [NC03]. However, 6,13-dihydropentacene $C_{22}H_{16}$ is expected to by a more stable pentacene derivative, which is why the authors suggest a free charge carrier induced reaction of a $C_{22}H_{16}$ into two $C_{22}H_{15}^+$ ions [Lan+04][NC03]. But the authors of [He+05] owe an explanation how this reaction is supposed to take place in a neutrally excited crystal as present also in our PL experiments, other than by autoionisation which usually requires larger excitation energies to be an efficient process [SW07]. Nevertheless, the energetic match is quite remarkable. Moreover, the emission has been attributed to a deep self trapped state without any further specification of its energetics or dynamics [Aok+01]. As discussed before for the emission band T, a self trapped state usually results from the polarization of the exciton's local environment making a second, energetically lower lying self trapped state of the same nature unlikely. However, the existence of trapped states at structural defects is not impeded by this restriction [PS99]. Indeed, in UPS studies trap states in the right energy range attributed to structural defects have been observed [Sue+09][Bus+13]. Bussolotti *et al.* [Bus+13] used N_2 and O_2 gas exposure to generate structural defects leading to energetic shifts between 0.1 eV to 0.2 eV. However, other authors do not observe a similar shift in the HOMO level upon gas exposure in their UPS experiments [Vol+06][Miz+17]. If the emission is related to structural defects, the occurrence should depend on the respective location of the measurement on the crystal surface. In the studies reported in this book, and in similar PL studies performed, a strong dependence on the excitation and detection spot cannot be confirmed within the spatial resolution of 1 μm. This excludes macro and mesoscopic defects such as twin domains or surface inhomogen-ities observed in the microscope images discussed above (*c.f.* figure 4.5). If a structural defect is responsible for the observed emission, it has to be of microscopic origin and of constant density at 1 μm^2 spatial resolution. Monomolecular step edges could pro-vide such a change in local environment, but the PFP cover layer is expected to quench the respective emission due energetically preferred CT formation as the molecular ori-entation at the steps resembles a cofacial geometry as present in PFP doped PEN thin films. However, a quenching of the DT_1 emission upon PFP coverage of the PEN (001) crystal surface is not observed. If a molecular impurity is the origin of the emission band, *e.g.* a hydropentacene species, this impurity should have a signature in the ab-sorption spectra. Indeed, in the thin film absorption discussed in section 4.2.2, be-low the assigned lower Davydov component of the $S_0 \rightarrow S_1$ transition, a non-zero ab-sorption can be found at roughly 1.7 eV which has been reported in other thin film samples as well [HHB80]. The DT_1 emission is red shifted by 210 meV with respect to the related absorption at 1.7 eV indicating a significant molecular reorganization energy. Under the assumption that the absorption and emission are caused by a $C_{22}H_{15}$ species this rather high value in energy can be rationalized by the non-planarity of the molecule which usually leads to significant molecular reorganisation during a HOMO-LUMO transition. The temperature dependence of the emission shows an overall decrease with rising temperature as shown in figure 4.11 upper right panel. The neat signal shows an inital slight increase (excluded data points, red, in brackets) before the overall signal decrease. This is attributed to interference with the DT_2 emission signal which shows a pronounced increase in the same temperature range. A fit by equation (4.4), as in case of the DT_2 emission intensity, was not possible further corroborating the peak-interference hypothesis. After excluding the data points attributed to interference with the DT_2 emission from the analysis a fit by equation 4.3 could be successfully performed and yielded an activation energy of (5 ± 2) meV corresponding well to an opitcal phonon mode with $7\,\text{meV}$ ($56.2\,\text{cm}^{-1}$) [Bri+02]. Moreover, in this energy range, a high phonon-charge carrier coupling has been reported [Mas+04] indicating an ionic molecular species such as $C_{22}H_{15}^{+}$ to be the origin of the emission band.

The energetically lowest emission component found consistently in every sample is located at $1.43\,\text{eV}$ and labeled DT_2. To the author's best knowledge it has not been reported in literature, so far. An energetically close lying peak between $1.35\,\text{eV}$ to $1.37\,\text{eV}$ has been related to a vibronic component of the DT_1 transition [Aok+01][Ang+12]. A similar component can be detected, especially for temperatures below $90\,\text{K}$, in the present PL signatures as well. Between $90\,\text{K}$ and $120\,\text{K}$ the intensity ratio of DT_1 and DT_2 differs between the neat PEN and the PFP/PEN sample. Furthermore, the opposite peak behavior has been observed for another PFP/PEN-PEN sample for which half of the PEN crystal substrate was covered with $20\,\text{nm}$ thick PFP layer. The observed intensity ratio between the PEN and the PFP/PEN measurements was inverted with respect to the intensity ratio reported above. This suggests a local effect of the substrate PEN crystal as the likely cause of the emission, as the characterizing parameters of the PFP cover layer do not change between samples. The intensity evolution with temperature differs significantly from that of the other three transitions. In particular a steep increase with rising temperature can be detected before the intensity drops again for even higher temperatures. To model this behavior, at first exclusively phenomenological, a second term is added to equation (4.3) to account for the intensity gain yielding

$$\bar{I}_{\text{int}} = A - B\exp\left(-\frac{E_a}{k_B T}\right) + C\exp\left(-\frac{E_b}{k_B T}\right). \tag{4.4}$$

Fitting of the two competing exponential functions with similar exponent by a global fit leads to unreliable results due to mutual inter-dependencies of the free parameters. Assuming an activation energy similar to that of the free exciton E and the DT_1 transition the first four data points at the low temperature side can be sufficiently described by the increase term $A + C\exp(-E_b/k_BT)$. Considering only these data points in the fit by the increase term leads to an activation energy for the intensity rise of $E_b \approx (9 \pm 2)$ meV. The resulting curve (dashed grey line) and its 95% confidence bands (striped grey area) are shown in figure 4.11 lower right panel. Using the obtained values for A, C and E_b as fixed parameters a global fit by equation (4.4) could be achieved (green curve) yielding a thermal activation energy for the non-radiative depopulation process of $E_a = (17 \pm 1)$ meV. The thermal activation at low temperatures together with the high locality identified for this transition suggests that the emission is the result of a trapped state immobilized at the surface or at crystal defects such as grain boundaries or dislocation lines. As the diffusive exciton transport is temperature activated, the state is continuously populated with rising temperature as more excitons are transported to the defects. Typical energies associated with exciton transport range from 10 meV to 20 meV in di-indeno-perylene [Top+14], to (25 ± 3) meV in rubrene [Gie+14] up to 55 meV in pyrene [Klö+72]. This places the obtained value for the DT_2 transition in PEN single crystals at the lower end of the energy range. The activation energy for the non-radiative decay channel again matches a PEN optical phonon mode with an energy of 17 meV (137.4 cm^{-1}) [Bri+02] suggesting a phonon mediated depopulation process. The proposed hypothesis explains the observed behavior reasonably well, however the mutual influence of DT_1 and DT_2 which is inherent to the analysis method might affect the obtained fit parameters.

Conclusively, the observed luminescence transition spectra are fully consistent with an exclusive PEN (001) single crystal emission which seems to be hardly influenced by the additional PFP cover layer. Even the singlet emission of PFP reported to occur between 1.71 eV [Ang+12][Rin+17] and 1.65 eV [Bro+17] could not be observed. This can be attributed to the much larger volume of the PEN crystal substrate, leading to a high luminescence intensity that masks the residual PFP luminescence signal. This, of course, could also hide a possible CT state contribution. For an unambiguous disclosure of the optical response of a head-to-tail PFP/PEN interface state more sensitive techniques have to be applied.

4.4.2. Interface Transitions: Differential Reflectance Spectroscopy

The CT state associated with the PEN:PFP complex can be identified by its absorption fingerprint around 1.6 eV as discussed previously in section 4.2.2. It has been reported

[Bro+11][Ham+20] that for mixed films, even at sub monolayer thicknesses, differential reflectance spectroscopy (DRS) can be successfully applied to retrieve the CT absorption signal for this D-A system. Furthermore, DRS is known to be highly sensitive to interface effects [BA92][Hos+10] and hence, can be used to examine the optical transitions at the PFP/PEN (001) single crystal interface. The measurements were performed according to section 3.4.4. An in-house written Matlab-script provided by Clemens Zeiser, University of Tübingen, was used to calculate the DRS and DDRS signals from the recorded reflectometry data. Data was binned in and averaged over 6 meV to 8 meV intervals to enhance the signal to noise ratio.

PEN crystals with their (001) surface facing up, were fixed on a glass slide with silver paint. The sample was cured for about 12 hours in a load lock section to avoid positional changes due to degassing effects by the silver paint. The sample position in the measurement chamber was calibrated by maximizing the PEN reflection in the white light reflection spectra. The correct positioning of the light spot was confirmed visually before each measurement to exclude probing of residual silver paint. The acquisition time and the number recorded spectra for averaging were chosen to yield a high signal to noise ratio, yielding a total recording time for each spectra of 6 s. Several hundreds of spectra were recorded and averaged to obtain the zero film thickness reference reflection spectra $R(0, \lambda)$. Thereafter, the evaporation of PFP proceeded at $0.4\,\text{Å}\,\text{min}^{-1}$.

Three distinct measurements were performed on samples treated with different annealing cycles, respectively labeled sample 1, 2, and 3. The first PEN crystal substrate, sample 1, was used as is. For sample 2, the PEN substrate crystal was thermally annealed in the measurement chamber at 80 °C for 1.5 hours and, subsequently, at 100 °C for 1 hour. Prior to PFP deposition and the optical measurement, it was cooled to room temperature within 3 hours. This Annealing was conducted to sublime residual water on the PEN crystal surface which could influence the interface interaction between PEN and PFP. For Sample 3, the PEN substrate crystal was kept at 90 °C during the PFP deposition to prevent adsorption of any residual water in the vacuum chamber.

The DRS signals of samples 1 to 3 are shown in figures 4.12 a), c) and e), respectively, for the first 42 Å of the PFP film, corresponding to about 2.7 monolayers. The overall signal can be divided into two energy regions. Below 1.8 eV, PEN is almost transparent (*c.f. section 4.2.2*) and hence, the signal is proportional to the extinction coefficient of emerging states related to the deposited cover film, as described by equation (3.19). Above 1.8 eV, equation (3.18) states that the signal is determined by the interplay of the dielectric functions of the PEN (001) crystal surface, of the deposited PFP layer and, potentially emerging states at the PEN-PFP interface.

4. Charge-Transfer States at Pentacene-Perfluoropentacene Interfaces

In the energy range below 1.8 eV, all DRS signals show two prominent features. First of all, at 1.78 eV an increase in the DRS signal is found scaling with the PFP film thickness. As this energy matches the lowest absorption component of PFP thin films (*c.f.* section 4.2.2) and as the signal strength is obviously correlated with the PFP layer thickness, this signal can be attributed to the low energy PFP transition.

At energies below 1.7 eV, the signal shows a steady decrease with thickness for all three samples, although the rate of decrease varies between the three samples. However, the mixed film CT absorption at ≈ 1.6 eV is absent in all spectra. To exclude a masking of the CT transition by the steady signal decrease at the low energy side, the DDRS signal has been calculated as the difference between two consecutive DRS signals. Assuming a steady negative offset with increasing film thickness, a correction for this offset could reveal potential hidden CT features at this energy range. Correcting for the averaged offset between 1.4 eV and 1.5 eV followed by a normalization to the change in the PFP $S_0 \to S_1$ absorption feature at 1.78 eV, the corrected $DDRS_c(t, E)$ signal at a film thickness t and energy E is calculated as

$$DDRS_c(t, E) = \frac{DDRS(t, E) - \overline{DDRS}(t, [1.4\,\text{eV}; 1.5\,\text{eV}])}{\overline{DDRS}(t, [1.75\,\text{eV}; 1.80\,\text{eV}])} \tag{4.5}$$

where $\overline{DDRS}(t, [E_1, E_2])$ is the arithmetic mean value of the respective DDRS signal at film thickness t between the energies E_1 and E_2. The energy range of the offset correction has been chosen to exclude interference with the potential CT absorption. The normalization is performed to enable comparability between samples. The $DDRS_c$ signal for all three samples is shown in figures 4.12 b), d), and f) for the first 20 Å, *i.e.* the first ≈ 1.3 monolayers. All three samples show the same features in the DDRS signal, *i.a.* zero amplitude between 1.4 eV and 1.7 eV all three samples implying the absence of any CT characteristics reported for thin film samples in the DRS spectra.

Even though the analysis of the various DRS and DDRS characteristics indicated no strong PEN-PFP interaction, *i.e.* the formation of a CT complex, in the low energy regime of the spectra, a contribution due to the pronounced PEN-PFP interaction cannot be excluded for energies above ≥ 1.8 eV. Therefore, comparison of the data with a simulated DRS signal can reveal the presence of additional spectral features not caused by Fresnel reflection of the two neat materials but by newly formed states at their common interface. The simulation of the DRS signal is, however, not trivial, as the spectrum for each film thickness depends on the complex dielectric function of the PFP layer $\hat{\varepsilon}_f$ and the PEN substrate crystal $\hat{\varepsilon}_s$ according to equation (3.18). As the PEN unit cell has a triclinic symmetry the dielectric function turns out to be very complex as shown by Dressel and coworkers [Dre+08]. Their results imply that even for a normal incident

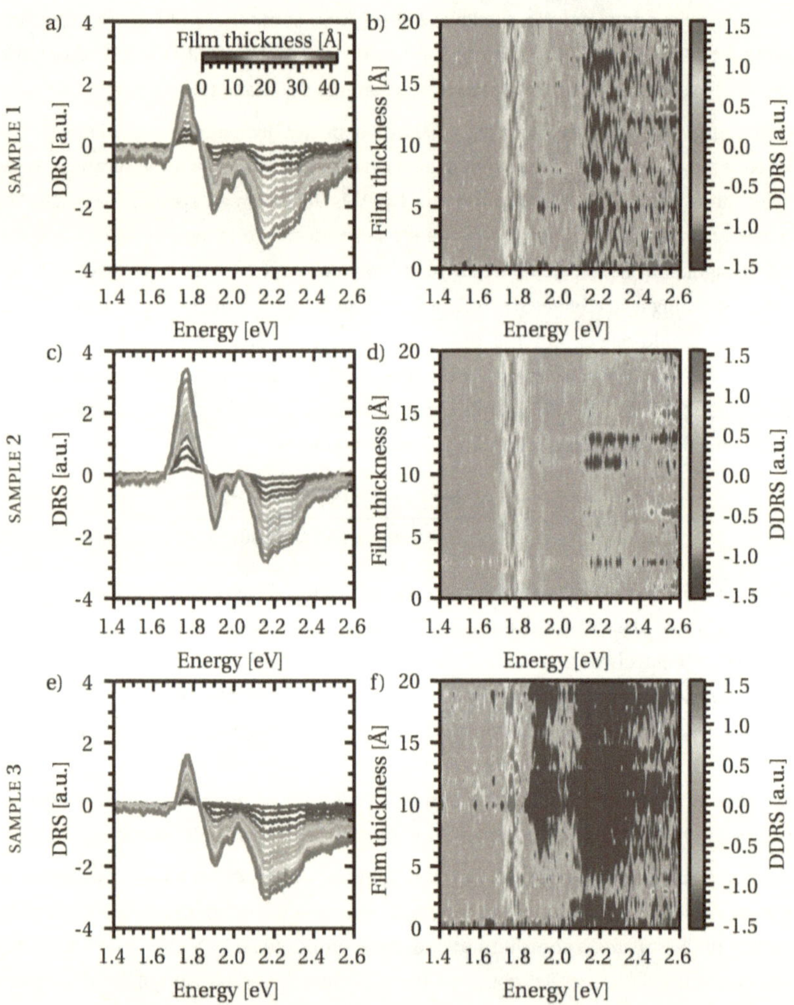

Figure 4.12.: a), c), and e) DRS signals and b), d), and f) DDRS signals of PFP thin film growth on a PEN (001) single crystal surface for different samples. Sample 1 (a) and b)): no further treatment. Sample 2 (c) and d)): annealed prior to PFP deposition. Sample 3 (e) and f)): annealed during PFP deposition. a) and b) adapted with permission from [Ham+20]. Copyright 2020 American Chemical Society.

light beam, the components of the dielectric function for each crystallographic axes $\hat{\epsilon}_{\vec{a}}$, $\hat{\epsilon}_{\vec{b}}$, and $\hat{\epsilon}_{\vec{c}}$ for the unit cell directions $\vec{a}$, $\vec{b}$, and $\vec{c}$, respectively, have to be considered in the reflection signal. Although, the main axis of the real and imaginary part of the dielectric function are not necessarily identical and might even rotate as a function of energy, as a first approximation a linear combination of the dielectric functions along the main crystallographic axis can be used

$$\hat{\epsilon}_s = a \cdot \hat{\epsilon}_{\vec{a}} + b \cdot \hat{\epsilon}_{\vec{b}} + c \cdot \hat{\epsilon}_{\vec{c}}. \tag{4.6}$$

A full coordinate transformation to an orthonormal lab system is not feasible due to uncertainties in the degree of polarization in the light source with respect the orientation of the $\vec{a}$ and $\vec{b}$ crystal axis of the substrate crystals. The data set published in [Dre+08] of the full dielectric function of crystalline pentacene between 1.5 eV and 2.5 eV has been generously provided by Dressl *et al.*. The refraction index and the extinction coefficient, which enables the calculation of the dielectric function of a PFP layer in the present thin film phase has been determined by Hinderhofer *et al.* [Hin+07]. As the unit cell of the PFP thin film phase is monoclinic with an angle $\beta = (90.4 \pm 0.1)°$ [Sal+08a] it is a reasonable approximation to only use the in-plane component of the dielectric function to model the normal incident reflection. Therefore, the out-of-plane data from Hinderhofer and coworkers [Hin+07] was manually digitized using the software OriginPro 2018b (ver. 9.55). All data has been interpolated using using the *InterpolatedUnivariateSpline-function* of the SciPy package for Python [Vir+20]. As a result, the dielectric function of PFP has been calculated from the refraction index and the extinction coefficient according to $\hat{\epsilon} = (n - i\kappa)^2$ [GM12].

The simulated DRS data is shown in figure 4.13 c) and compared to the DRS data of sample 1 depicted in the subfigure a). The parameters a, b and c chosen to model the dielectric function described by equation (4.6) of the PEN substrate crystal are 0.8, 0.05, and 0.15, respectively, and have been determined by iterative optimization. The grey areas in figures 4.13 c) and d) indicate the boundaries of the simulated spectra due to a lack of data on the dielectric function outside of the given energy range. An important result is the signal decrease on the low energy side turns out to be an inherent property of the two medium system and not an extrinsic effect caused, for example, by an enhanced diffuse surface scattering due to the formation of PFP agglomerates on top of the PEN crystal. Furthermore, it is evident that the simulated spectra show the same characteristics as the measured data (marked by dotted lines), even though the absolute intensities are different. Comparing the energies of the occuring peaks yields a coincidence between data and simulations within 10 meV, which is in good agreement taking into account the binning range. A DDRS map calculated from the

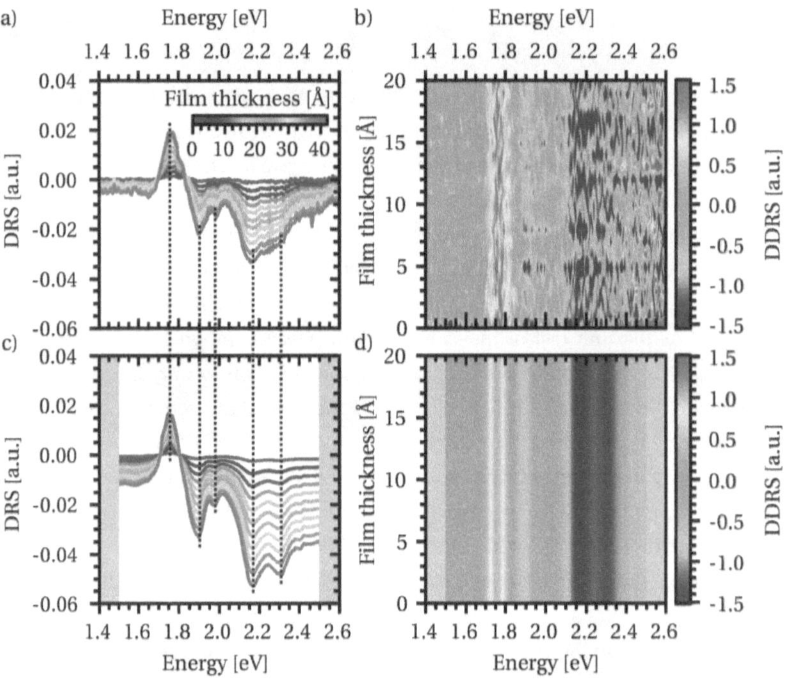

Figure 4.13.: a) and b) show DRS and DDRS signal of the untreated sample 1 from figure 4.12 a) and b). c) and d) are simulated from PFP and PEN data. The excellent agreement of the spectral features (dotted lines) of simulation and data further corroborates the needlessness of an inclusion of a CT transition to explain the observed data. Not simulated regions are depicted as grey areas. This graphic has been previously published alongside [Ham+20] as supplementary information.

simulated DRS signal is shown in figure 4.13 d) while the corresponding DDRS data of sample 1 is depicted in subfigure b). The simulated DDRS signal is offset corrected by $\overline{DDRS}(t, [1.5\,\mathrm{eV}, 1.6\,\mathrm{eV}])$ and normalized to $\overline{DDRS}(t, [1.74\,\mathrm{eV}, 1.78\,\mathrm{eV}])$ to comply with the procedure applied to the experimental data. As can be seen, the experimental DDRS is well reproduced by the simulated data.

While the spectral signatures of the DRS signal and the DDRS maps are remarkably well reproduced by the simulating the reflection PEN-PFP system, the deviation in intensities between experimental data and simulations, but also among the three different samples requests further explanation. First of all, the DRS signal is highly sensitive to changes in the incident angle of the light [KWS97] and equation (3.18) used for modeling is strictly valid only for normal incidence [MA71]. Two experimental constraints can lead to deviations from perfect normal incidence. The crystals show a macroscopic surface roughness (*c.f.* figure 4.5). Even though the light beam shows a perpendicular incidence with respect to the macroscopic crystal surface, individual grains can be tilted with respect to each other leading to a slightly varying incidence angle on microscopic length scales. Additionally the mounting by silver paint bears the risk of slightly tilting the surface of the single crystal sample with respect to the sample stage. As the available tilting stage in the vacuum chamber only allows for one rotational degree of freedom, the light incidence can be optimized to 90° only in one dimension. Secondly, the theoretical DRS signal derived by McIntyre and Aspnes [MA71] is based on the assumption of growth of a uniformly closed film, which is not met by the Vollmer-Weber growth present for PFP deposited on the PEN surface. The uncertainties in the incidence angle combined with the non-perfect film growth can, at least qualitatively, explain the observed deviations in the DRS signals among the three samples and the intensity differences compared to the simulation.

In summary, the DRS and DDRS signals of the three samples can be consistently explained by assuming a two layer system composed of a PEN (001) crystal substrate substrate with a closed PFP layer on top under normal incidence of light. According to the agreement between experimental data and simulations, there is no need to assume any additional electronic transitions, apart from the ones in the neat materials PEN and PFP, to explain the observed DRS signal. Under the constraints discussed above the scientific principle of preferring the model containing the least number of necessary assumptions (*Occam's Razor*) [Nov+19] leads to the conclusion that no additional inter molecular electronic transition related to the PFP/PEN head-to-tail dimer is present at the PFP/PEN (001) interface.

4.4.3. CT Formation and Molecular Arrangement

To fully explain the missing CT or even dimer state signature in the PL and DRS measurements, quantum chemistry calculations based on the dimer approach[EE17] were performed by Marian Deutsch and Bernd Engels from the Institute of Physical and Theoretical Chemistry at the Julius Maximilians University Würzburg. The results have been published in a collaborative publication [Ham+20].

The calculations included two different dimer configurations for the neat PEN and the neat PFP unit cell as a reference system, each of which based on the published crystal structure for PEN [Sie+07][2] and for PFP [Sak+04]. Furthermore, the cofacial arrangement for a mixed PEN-PFP dimer has been realized using the inter molecular distance proposed by Kim *et al.* [Kim+19] as well as a geometrical optimized geometry in which the molecules are shifted $0.7\,\text{Å}$ along the short molecular axis. A head-to-tail orientation of PEN and PFP was achieved by stacking both molecules along their long axis and optimizing their inter molecular distance in $0.1\,\text{Å}$ steps revealing an equilibrium distance of $2.62\,\text{Å}$.

Based on these optimized geometries, vertical excitation energies were calculated for various molecular geometries in the excited state, explicitly for the case that both molecules' atoms are in the electronic ground state geometry after excitation (S_0/S_0), either PFP or PEN relaxes to its excited state geometry (S_1/S_0 and S_0/S_1, respectively), or PEN adapts its cationic and PFP its anionic ground state geometry (+/-). The latter geometry was explicitly chosen as it is supposed to favor CT formation. The respective vertical excitation energies as well as the CT character, *i.e.* a value between 0 and 1 characterizing the partial charge transfer with 1 for a full charge separation and 0 for a neutral dimer state, of the different hetero dimer orientations and excited state geometries are listed in table 4.5 [Ham+20].

These calculations yield a clear picture of the inter molecular electronic processes on a molecular scale and are in full accordance with the experimental evidences. For the neat PEN and PFP homo dimers, the vertical excitation energies to the first excited state depend on the respective molecular orientation at the excited state geometry. For the diabatic excitation, *i.e.* S_0/S_0 excited state geometry, the excitation energies for the PEN and PFP dimer lie between $2.35\,\text{eV}$ and $2.36\,\text{eV}$ and between $2.08\,\text{eV}$ and $2.20\,\text{eV}$, respectively. In the case of an adiabatic excitation, *i.e.* S_0/S_1 excited state geometry, the excitation energies are overall lower and lie between $2.14\,\text{eV}$ and $2.15\,\text{eV}$ and between $1.90\,\text{eV}$ and $1.97\,\text{eV}$ for PEN and PFP, respectively. For the PEN-PFP hetero dimer in the cofacial geometries the excitation of the first excited state is found to

---

[2]Structure accessible at Cambridge Crystallographic Data Center - CCDC - with deposition number 619979.

Dimer	State	S_0/S_0		S_0/S_1		S_1/S_0		+/-	
		E [eV]	CT	E [eV]	CT	E [eV]	CT	E [eV]	CT
cofacial	S_1	1.87	0.42	1.75	0.31	1.83	0.47	1.73	0.60
	S_2	2.35	0.08	2.30	0.43	2.36	0.07	2.25	0.14
	S_3	2.57	0.65	2.65	0.43	2.52	0.71	2.44	0.42
	S_4	3.07	0.24	3.12	0.72	3.11	0.73	3.16	0.47
cofacial shift	S_1	1.95	0.38	1.81	0.26	1.89	0.44	1.80	0.57
	S_2	2.31	0.19	2.26	0.63	2.37	0.08	2.21	0.25
	S_3	2.57	0.54	2.67	0.30	2.49	0.68	2.44	0.27
	S_4	3.14	0.16	3.10	0.80	3.10	0.80	3.16	0.32
head-to-tail	S_1	2.23	0.00	1.97	0.00	2.17	0.00	2.16	0.00
	S_2	2.40	0.00	2.63	0.00	2.43	0.00	2.36	0.00
	S_3	3.23	0.00	3.31	0.00	3.38	1.00	3.17	1.00
	S_4	3.40	0.10	3.37	0.00	3.43	0.00	3.35	0.00
	S_5	3.40	0.90	3.39	1.00				

Table 4.5.: Vertical excitation energies and CT character of the PEN-PFP hetero dimer for the different dimer orientations "cofacial", "cofacial shifted", and "head-to-tail" and for different excited state geometries. Values reproduced from [Ham+20]

be significantly lower in energy than the respective diabatic and adiabatic transitions in the neat dimers (*c.f.* table 4.5). While for the neat dimers, the CT contribution is low ($CT_{PEN} \leq 0.11$ and $CT_{PFP} \leq 0.18$) and for the diabatic transition the states localize on one monomer, the cofacial dimers show a non vanishing CT character in the first excited state and a delocalization over both monomers. In line with previous studies [Bro+11][Kim+19], this transition is assigned to the 1.6 eV CT band found in mixed films.

For the head-to-tail orientation, the calculations reveal the states to be localized on one of the monomers. The vertical excitation energies are close to the adiabatic transitions of the neat PEN or PFP homo dimers, depending on which of the hetero dimer's molecules relaxes to its excited state geometry. Either the PEN molecule relaxes to its S_1 geometry yielding the S_1/S_0 configuration which results in an excitation energy close to the neat PEN dimer or, *vice versa*, the PFP molecule relaxes which leads to the S_0/S_1 excited state geometry and yields a vertical excitation energy coinciding to the on of the PFP homo dimer. Remarkably, if the hetero dimer is in its +/- excited state geometry the first excited state shows neither CT character nor delocalization. The first CT state in this particular dimer orientation is found to be the S_5 state for the excited dimer to be in its respective ground state geometry (S_0/S_0) or for the PFP molecule in excited state geometry (S_0/S_1), and the S_3 state for the PEN molecule in its excited state geometry (S_1/S_0) or for the dimer molecules in their respective ionic molecular geometry

(+/-). For either case, the CT state is about 1 eV higher in energy than the respective first excited state.

These theoretical findings support the conclusion drawn from the DRS measurements and the comparison with the simulated DRS signal. No strong excited state interaction seems to be present at the standing-stack hetero interface due to the present head-to-tail geometry and the resulting low binding energy of any possible CT state. This in return, is also expected to influence the formation of a charge transfer state by free charge carriers at these interfaces. While in the mixed layer, charge carrier pairs can relax into lower excited states of the cofacial dimer, the CT state in the head-to-tail geometry is energetically unfavorable, leading to only slightly bound CT states at the interface [Ham+20]. The consequence of this observation on the optoelectronic properties of PEN-PFP heterojunctions are discussed in the next section.

4.5. Interface Engineering in D-A devices

The formation of CT states as well as their binding energy, their relaxation, and subsequent separation into free charge carriers strongly influence the performance of any D-A device, *c.f.* for example [Opi+09][Van16]. As shown in the previous sections, the molecular arrangement at the PEN-PFP interface enables the control of the occurring interface states. To show the potential application of controlling charge transfer formation by guiding molecular orientation via interface engineering, two prototypical device architectures will be compared in the following. A mixed PEN:PFP layer with a cofacial D-A orientation and a PFP/PEN bilayer with a head-to-tail orientation define the references geometries of the subsequent analysis.

The results and their interpretation presented in the following section have been published in [Ham+20] if not stated otherwise.

4.5.1. Device Architecture: Structural and Optical Characterization

Schemes of the device architectures are shown in figure 4.14. Both device types comprise a transparent ITO electrode covered by a spin coated PEDOT:PSS film on top as hole transport layer and to smooth the ITO surface [PYF03]. For the bilayer (BL) device in figure 4.14 a), the electrode is followed by a 30 nm thick PEN layer and a subsequent 30 nm thick PFP layer, both deposited by vacuum sublimation at a deposition rate of $0.17\,\text{Ås}^{-1}$. The device depicted in figure 4.14 b) comprises a 30 nm thick bulk hetero junction (BHJ) with a molecular PFP:PEN ratio of approximately 1:3 prepared by simultaneous evaporation from separated crucibles. The BHJ layer is sandwiched between a PEN and a PFP layer of 15 nm thickness each for hole and electron transport, respectively. Prior to the deposition of the 80 nm thick Ag contacts, both device setups have been covered by a 5 nm thick BPhen layer to counteract metal atom penetration [Sch+05] while, at the same time, allowing for maximal device performance [Ste+12]. On each substrate a total of 14 circular diodes is prepared through a shadow mask, yielding seven 0.5 mm and 1.5 mm diameter devices, respectively. After contact deposition, the devices are encapsulated under nitrogen atmosphere to enhance their durability. The circular zoom-in for both device schemes shows the expected charge carrier behavior at the respective D-A interface in a simplified, yet essential manner. While in the BL device, electrons and holes are expected not to form a tightly bound CT pair, the molecular orientation in the BHJ device leads to a capturing of charge carriers of opposite polarity in a bound CT state, effectively binding both free charge carriers. To characterize the PEN-PFP interface in the BL device, a 40 nm PEN film has been deposited on an otherwise similarly treated PEDOT:PSS/ITO substrate as reference.

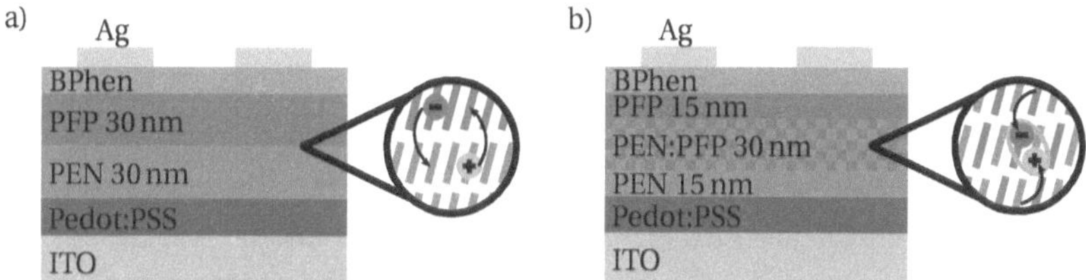

Figure 4.14.: Device architecture of a) the PFP/PEN bilayer device and b) the PFP:PEN bulk hetero junction device. Circular zoom-ins show schematically the charge carrier interaction at the respective interface. Reprinted with permission from [Ham+20]. Copyright 2020 American Chemical Society.

After optoelectronic characterization discussed in the following section, the devices' encapsulation was removed to measure their UV/Vis absorption as well as their integral structural characteristics via XRD. The results of these examinations are shown in figure 4.15.

The small angle diffraction patterns of both devices as well as of the reference sample are displayed in figure 4.15 a). All three samples show a first order Bragg reflection at around $q_z = 0.4\,\text{Å}^{-1}$. An XRR analysis of the small angle diffraction pattern leads to two significant results. First of all, all three samples show superimposed Kiessig fringes. According to equation (3.10), the periodicity obtained by FFT analysis of $\Delta 1/q \approx (240 \pm 15)\,\text{Å}$ yields a film thickness of (150 ± 10) nm. This well relates to the ITO thickness on the glass substrate (*c.f.* figure 4.16 upper left panel). The BHJ device's overall intensity and its Kiessig fringes appear to be damped compared to the BL device. This indicates, according to equation (3.9), a higher interface roughness at the consecutive organic interfaces. This result agrees with an expected higher disorder due to phase separation [Hin+11] in the mixed layer of the BHJ compared to the stacked layers at the PEN and PFP interface in the BL sample.

The samples' second order Bragg reflections depicted in figure 4.15 b) show a clear distinction between the samples' out-of-plane lattice constants. The reflection peak at $q_z = 0.812 \text{Å}^{-1}$ of the reference sample corresponds to a lattice constant of $d_{001} = 15.5 \text{Å}$ in good agreement with the PEN thin film phase [Mat+03b]. For the BHJ device, the single reflection peak at $q_z = 0.800 \text{Å}^{-1}$, transferring to a lattice distance of $d_{100} = 15.7 \text{Å}$, is in good agreement with the PFP thin film phase's (100) lattice spacing [Sal+08a]. These XRD results, suggest a highly crystalline PFP layer on top of a disordered PEN:PFP mixed layer corroborating the hypothesis of a mixed layer comprising small, phase separated PEN and PFP crystallites yielding only a weak diffraction signal. The diffraction signal of the BL device can be decomposed into two peaks centered at 0.801Å^{-1} and 0.812Å^{-1} which are expected for a PEN thin film covered by a PFP layer and corroborating the molecular standing stack arrangement at the interface. No additional reflection signals were detected.

In figure 4.15 c) and d) the UV/Vis absorption of the two device types is shown. The absorption has been calculated according to equation (3.15). The measured transmission T_m has been corrected for the PEDOT:PSS/ITO substrate transmission T_S, measured on a reference sample. With $T_D = T_m/T_S$ the device transmission T_D is obtained. The reflectance at the organic/air interface of the device has been corrected for additional back reflectance at the glass/air interface of the ITO substrate. As the back reflectance is not directly accessible, it is approximated by the reflectance of the substrate/air interface R_S of a reference PEDOT:PSS/ITO sample. The measured reflectance R_m is given by the reflectance at the device's air/organic interface R_D and the back reflection modulated by the back and forth transmission of light through the organic layers:

$$R_m = R_D + T_D^2 \cdot R_S. \tag{4.7}$$

The results obtained by this correction have to be carefully evaluated for each sample. In the present case, the correction enables the identification of individual transitions, however, a quantitative analysis of relative transition strengths cannot be conducted. A comparison with the absorption transition spectra of PEN and PFP thin films in section 4.2.2 yields the following conclusions. The absorption peaks at 1.86 eV and 1.96 eV result from a Davydov splitting of the lowest PEN transition while the shoulder at 1.77 eV can be attributed to three distinct PFP transitions. The clear expression of the PEN

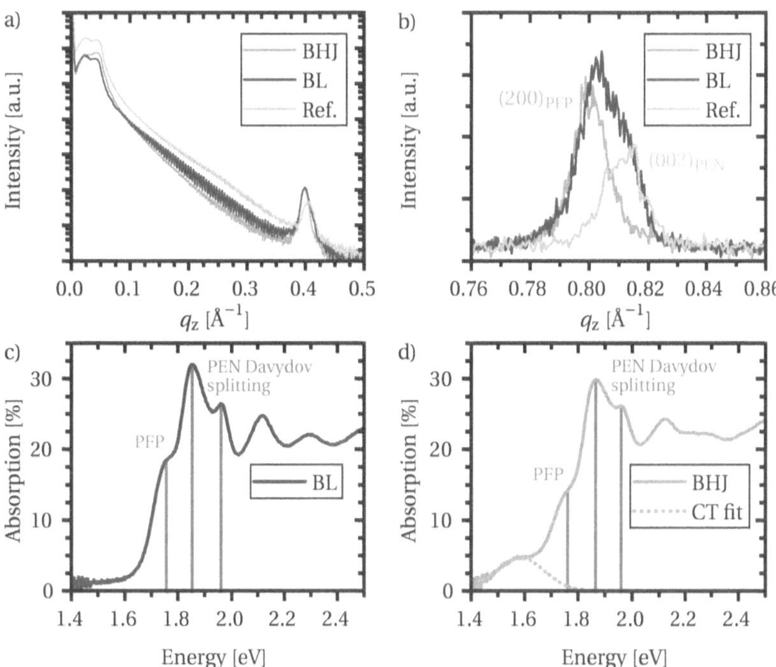

Figure 4.15.: a) Small angle X-ray diffraction patterns well as b) second order Bragg reflection of bilayer (BL) device, bulk heterojunction (BHJ) device, and reference sample. UV/Vis absorption of BL device c) and BHJ device d) including the estimated CT absorption. Adapted from data previously published as supplementary information in [Ham+20].

Davydov splitting in the absorption spectra of the BHJ device confirms the existence of a local crystalline PEN phase which could not be detected by XRD. The overall high absorption background at the related PEN vibronic progressions above 2 eV could either be caused by an underlying broad PFP absorption, as observed for neat thin films, or be an artifact of the reflection correction. The slight blurring of the features in the BHJ device is attributed to an overall lower crystallinity, especially in the mixed PEN:PFP layer, and the concomitant increase in static part of the energetic disorder. Despite the broad high energy absorption spectra, a clear difference between the optical characteristics of both device types becomes evident at around 1.6 eV. For the BHJ an absorption feature which is missing in the BL device is found. The feature can be approximated by a Gaussian peak centered at 1.58 eV with a FWHM of 200 meV and hence, can be attributed to the CT state formed as a consequence of the cofacial molecular arrangement. This finding confirms the much higher CT density in the BHJ device upon illumination compared to the BL device.

The spatial configuration of the PFP/PEN interface in the BL device is vital for its performance. The molecules are thought to form a standing stack at the heterojunction, however, a rough surface will lead to additional cofacial molecular arrangements and hence, diminish the out performance of the BL versus that of the BHJ device. In figure 4.16 the two upper right panels show an AFM image of the PEN (40 nm)/PEDOT:PSS/ITO/glass reference sample. The ITO-covered glass substrate comprises an about 1 cm wide ITO stripe on top of the glass substrate defining the active area of the devices and leaving two stripes of neat glass on both sides for electronic contacts. The AFM height data shown here is recorded at the ITO/glass edge. The lower lying plane (left, blue colored) is the PEN/PEDOT:PSS covered neat glass substrate while the step and the terrace thereafter is the result of the ITO cover layer on the glass. The profile line at position A has been fitted with a step function by the Gwyddion software package yielding a step height of (131 ± 7) nm which is in good agreement with the value deduced from the Kiessig fringes. The zoom-in shows a region of interest on the ITO side of the substrate, where large crystalline PEN domains of µm size grow on the PEDOT:PSS layer. The morphology of the crystallites corresponds to the Vollmer-Weber island growth with the growth of the individual grains beeing promoted by double screw dislocations leading to mono molecular steps on top of the crystallites [Rui+04]. The data of four representative profile lines indicated in the AFM image are depicted in the lower subgraphs of figure 4.16. The (001) surface of the crystallites is interrupted by step edges with heights corresponding to the PEN (001) lattice spacing of 15.4 Å [Mat+03b]. This substantiates the claimed standing-stack geometry for the PEN

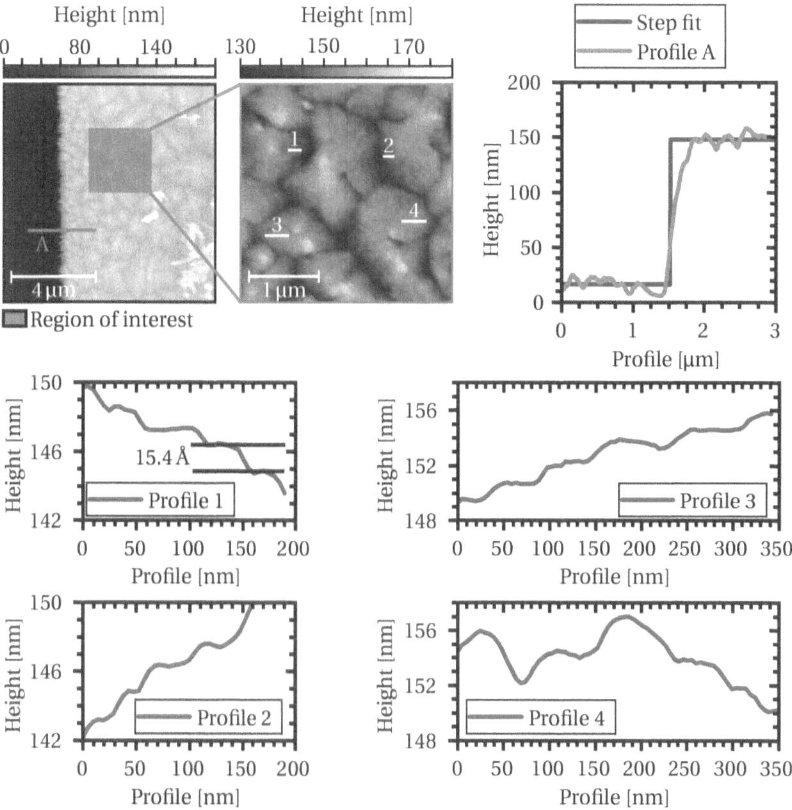

Figure 4.16.: AFM study of the pentacene thin film morphology on top of a PEDOT:PSS/ITO covered glass substrate. Upper left panel shows a 40 nm thick PEN film grown on PEDOT:PSS/ITO/glass substrate at the ITO/glass step edge. Zoom-in shows a detailed view on the PEN domain morphology established on the PEDOT:PSS/ITO covered part of the glass slide. Right upper panel depicts the profile line A measured across the ITO/glass step edge. Profiles 1 to 4 corroborate the monomolecular steps on top of the individual PEN crystallites. Adapted from the supplementary information in [Ham+20].

template underneath the PFP coverlayer, while PFP growth studies on the PEN (001) single crystal surfaces suggest a predominant head-to-tail orientation at the interface.

4.5.2. Controlling Charge Transfer Trap States

To examine the influence of a charge transfer trapping mechanism on the overall device performance, as a precondition, ambipolar charge injection has to be established. Conclusive evidence for the presence of both charge carriers, electrons and holes, is their radiative recombination, *i.e.* the occurrence of electroluminescence (EL). For both device types the integrated EL intensity normalized to its maximum as a function of the applied voltage is shown in figure 4.17 a). As both devices have a very low quantum efficiency acquisition times of 60 s were necessary to record EL spectra of sufficient signal to noise ratio. The voltage steps were chosen to $\Delta V = 0.5\,V$ and the applied voltage was limited to 4.5 V. These limits were chosen to minimize thermal stress caused by Joule heating, as operation at elevated temperatures has repeatedly been identified in many studies to cause accelerated device degradation [She+96][IT02][Tya+16].

The first major result of these studies is the occurrence of EL and hence, the confirmation of ambipolar charge injection and transport. Estimating the EL onset by means of a linear function to describe the increase with the applied voltage yields 3.0 V for the BL and a slightly higher onset at 3.5 V for the BHJ device. Both values are higher than the built-in voltage of 0.9 V caused by the energetic offset between the PEDOT:PSS anode and the Ag cathode indicating the presence of additional interface effects (*c.f.* section 2.4.1). A lower electron injection barrier would have been expected for Ca/Al contacts. However, devices based on such cathode materials did not show a significant current through the device. It is reasonable to suspect the formation of fluorite (CaF_2) during the Ca deposition onto the PFP ($F_{14}C_{22}$) layer as a defluorination of the PFP molecules and the subsequent formation of metal-fluorides has been observed on Cu (111) surfaces [Sch+12]. This, however, could not be experimentally confirmed with the conducted optoelectronical studies. As fluorite is highly insulating [Sut+08], the observed device behavior could be expected. The 0.5 V difference in the EL onset of both device types can be associated with two effects. At first, the structural studies on both devices indicate an overall lower crystallinity for the BHJ diode, presumably caused by the mixed PEN:PFP layer. Furthermore, the intermixing of PEN and PFP molecules leads to the multiple occurrence of energetic barriers hindering the transport of the individual charge carriers. The lower lying HOMO level of PFP imposes an energetic barrier for hole transport in the HOMO levels of PEN and, *vice versa*, the higher lying PEN LUMO acts as a barrier for an electron transport in the PFP LUMOs. Both effects lead to an overall reduction of the charge carrier mobility in the mixed layer of the BHJ diode, finally yielding an accumulation of charge carries with opposite polarity at the mixed layer boundaries. The resulting electric field partially compensates for the applied field and has to be overcome for charge carrier injection into the mixed layer.

The areal charge density necessary to generate the $U = 0.5\,\text{V}$ potential difference across the $d = 30\,\text{nm}$ thick PEN:PFP mixed layer with a mean dielectric constant of $\epsilon_r \approx 3$ can be estimated as

$$\frac{Q}{A} = \epsilon_r \epsilon_0 \frac{U}{d} \approx 2.8 \cdot 10^{-3} \frac{e}{\text{nm}^2}. \tag{4.8}$$

Accounting for the two molecules per unit cell and the thin film phase unit cell parameters of PEN ($a = 5.77\,\text{Å}$, $b = 7.49\,\text{Å}$) leads to less than one charge carrier per 1000 molecules at the boundaries between the central blended layer and the adjacent electron and hole transport layers[3].

The spectrally resolved EL at 4.5 V normalized to the emission maximum of both device types is shown figure 4.17 b). For both diodes, the emission is centered around 1.4 eV which can be associated with the CT emission of a cofacially stacked PEN-PFP molecular pair [Ang+12]. This is expected for the BHJ diode, but, at first glance, seems counter intuitive for the BL device. However, as previously discussed, the PFP/PEN interface in the BL device comprises a large number of molecular step edges where a cofacial arrangement of PEN and PFP molecules can occur. The BL emission shows a substantial contribution at the high energy side of the spectrum. As the EL of a neat PEN sample is dominated by the self trapped state around 1.65 eV [SMS09] this could account for the high energy contribution, however, no spectral feature can be ascribed to this high energy emission band. Most recently, the emission of inter molecular charge transfer states was consistently described in a Franck-Condon picture with inter molecular vibrations [TBV20], similar to an excimeric emission (*c.f.* section 2.3.2). In this picture, the detected emission is the result of a radiative decay into high levels of inter molecular vibrations of the electronic ground state causing a broad emission band. The spectral signature of the BHJ emission shows a much narrower profile compared to the broad BL signal. Accounting for the high density of cofacial PEN-PFP dimers in the mixed layer, the resulting self absorption of the CT state (orange dotted line in figure 4.17 b) can explain the spectral shape. Correcting the BL emission for CT state self absorption with the fitted Gaussian function derived from the absorption spectra and, in particular, its CT band in the previous section, the BHJ spectrum can be reproduced (blue dots figure 4.17 b). This leads to the conclusion that in both device types, the cofacial CT state is the main luminescent trapping mechanism for electron-hole pairs.

Comparing the overall intensity, the BL diode shows an over five times higher integrated intensity than the BHJ device, in contradiction to the expected higher CT state density in the BHJ device. In the BL device, the electrons and holes are transported mainly undisturbed to the PFP/PEN interface, where, before passing into the adjacent

---

[3]For the unit cell parameters ($b = 4.51\,\text{Å}$, $c = 11.48\,\text{Å}$) of PFP with two molecules per unit cell [Sal+08a], the estimation of less than one charge carrier per 1000 molecules at the interface remains unchanged.

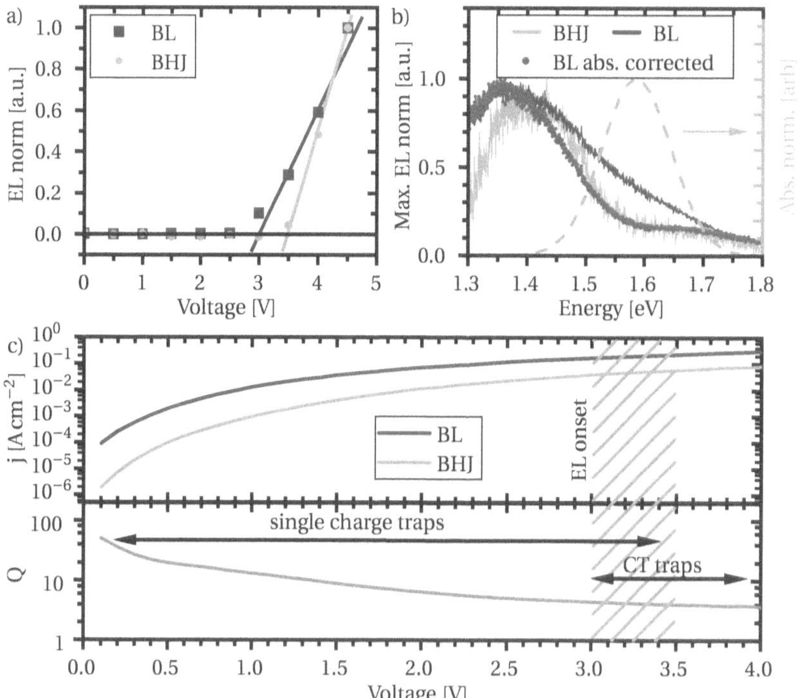

Figure 4.17.: a) Integrated electroluminescence (EL) normalized to the maximum intensity at 4.5 eV with linear approximations of EL onset. b) Spectrally resolved EL of both devices, as well as self absorption corrected BL device emission. The CT absorption of the cofacial hetero dimer determined by the UV/Vis absorption in figure 4.15 d) is shown in orange with respect to the right y-axis. c) Average current density - voltage characteristics of BL ($N = 5$) and BHJ ($N = 4$) devices (upper panel) as well as the related Q-factor defined by the ratio of both current densities at a given voltage. The dashed orange region marks EL onset. Reproduced with permission from [Ham+20]. Copyright 2020 American Chemical Society.

layer, mostly intermediate quasi free or weakly bound CT pairs are formed by virtue of the molecular head-to-tail arrangement. Some charge carrier pairs will be trapped in the cofacial arrangements at the PEN step edges. Assuming a Langevin trapping mechanism, the Coloumb capturing radius at room temperature is given by $r_c = 19\,\mathrm{nm}$ for $\epsilon_r \approx 3$ according to equation (2.67). The device characteristics displayed in figure 4.17 c), yield a current density of $j \approx 0.3\,\mathrm{Acm}^{-2}$ at 4 V which results in an average time of a CT state occupation of

$$\tau = \frac{e}{\pi r_c^2 j} \approx 47\,\mathrm{ns}\,. \tag{4.9}$$

This is much longer than the life time of a PEN:PFP cofacial CT state which has been determined to be $\leq 1\,\mathrm{ns}$ [Bro+17], showing that no emission saturation is expected. For the BHJ diode the charge transport into the mixed layer, where the CT formation occurs, is hindered by accumulated charge carriers at the boundaries as discussed before. At 4 V, the current density in the device amounts to $j \approx 0.1\,\mathrm{Acm}^{-2}$ according to figure 4.17 c). With equation (4.9) this yields an average CT state occupation time of $\tau \approx 140\,\mathrm{ns}$ explaining the overall lower luminescence for the same applied voltage.

In figure 4.17 c) (upper panel) the averaged current density - voltage characteristic for the BL and BHJ devices is shown as a function of the applied voltage demonstrating an overall higher current density for the BL diode. In total, $N = 5$ BL diodes and $N = 4$ BHJ diodes have been measured. As the difference between the average current densities of the two device types is larger than the statistical standard error, the error bands are omitted for clarity. To quantify this enhancement the current density ratio

$$Q = \frac{j_{\mathrm{BL}}}{j_{\mathrm{BHJ}}} \tag{4.10}$$

is plotted in the lower panel of 4.17 c). For low voltages, a high value of $Q \geq 10$ is found, which decreases with increasing voltage. This advantage of a better electronic performance of the BL device is associated with a lower density of shallow single charge carrier traps resulting from the better structural integrity as well as the absence of HOMO and LUMO mismatches as in case of the BHJ devices. With increasing voltage and current density the trapped charge carriers can be released by a field induced Poole-Frenkel mechanism [SW07]. Above the EL onset, Q levels at a value of approximately 4, caused by the decreasing contribution of shallow traps with increasing voltage and the CT charge carrier pair trapping mechanism being now dominant in this operation regime of the device. As the density of cofacial PEN-PFP dimer pairs is much higher in the mixed layer of the BHJ its device performance is consistently lower. The high contribution of CT trap states in the BHJ diodes is again corroborated by the integrated EL intensity shown in figure 4.17 a). After the EL onset in the BHJ the slope of the EL intensity increase is larger for the BHJ device, indicating an easier accessibility of the emissive CT states.

The examination of these prototypical devices shows that the molecular arrangement at the D-A interface substantially influences the charge transport properties as well as the optoelectronic device properties. This promises tailoring and optimization of organic heterojunction properties in thin film applications by controlling and tuning the interfacial molecular arrangement without need for additional chemical modifications.

5. Interplay of Crystal Structure and Electronic Transitions in Zinc Phthalocyanine Aggregates

5.1. Crystal Structure and Excited States

In crystalline aggregates the molecules are positioned in a regular and periodic manner and hence, their respective orientation is not random but underlies the same symmetry conditions. Depending on the degree of ordering these constrains can range from the nanometer scale, *i.e.* a few lattice constants, to macroscopic dimension, meaning micrometer or even millimeter scales. The molecular orientation related to the crystal structure and the crystallinity crucially influences the photophysics of the molecular aggregate. As outlined in section 2.3.4, the molecular orientation determines the absorption and emission energetics and dynamics, *e.g.* by inducing J- or H-aggregate excitons [GP13] or by a Davydov splitting of the excited states. The number of coherently coupled monomers strongly influences the emission intensity and its spectral distribution [Zha+16] as does the electronic coupling strength [Spa10]. The molecular packing is also known to influence singlet fission [Wan+14][PB15][Ari+16] and can lead to self trapping of excited states as previously discussed for pentacene in section 4.4.1. Furthermore, the underlying crystal structure can determine the excited state character, as it is the case for perylene for which the respective polymorph energetically selects the formation of either an excimer or a monomeric excited state[TKT74].

Zinc phthalocyanine (ZnPc) is a widely used material with applications ranging from photo assisted cancer therapy to fundamental research, however, the nature of its electronic transitions is still under discussion in literature (*c.f.* section 3.1.2). Although it has been widely studied for photovolatic applications [Koe+05][Bru+10][Mei+12][Ike+14][Bre+15], the underlying microscopic photophysical processes have not been clarified to the same extent [Yos+73][Bał+09].

In this chapter, the temperature dependent emission of the α and β phase of ZnPc is characterized, using crystalline thin films and single crystals and the occurring photophysical processes are related to the respective crystal structure. The $\alpha \rightarrow \beta$ phase transition kinetics are disclosed, leading to the application of this material in dual luminescent OLEDs, published in [Ham+19].

5.2. Solvent Spectra

To gain a complete picture of the complex photophysical processes governing the electronic excitations of ZnPc aggregates, the compound has to be studied in relation to its monomer. For this purpose, the photon absorption and emission of ZnPc solutions of various concentration are characterized.

The poor solubility of ZnPc in many non polar solvents [NGS87] impedes a systematic examination by concentration dependent studies. According to Ghani *et al.* [GKR12] the highest solubilities for conventional solvents are found for dimethylacetamide (DMAC, $c = 9.96 \cdot 10^{-4}\,\mathrm{mol\,kg^{-1}}$), dimethyl sulfoxide (DMSO, $c = 5.58 \cdot 10^{-4}\,\mathrm{mol\,kg^{-1}}$), N-methyl-2-pyrrolidone (NMP, $c = 6.92 \cdot 10^{-3}\,\mathrm{mol\,kg^{-1}}$) and tetrahydrofuran (THF, $c = 7.79 \cdot 10^{-4}\,\mathrm{mol\,kg^{-1}}$) as well as for the inoic liquid 1-ethyl-3-methylimidazolium acetate (STOR, $c = 9.19 \cdot 10^{-4}\,\mathrm{mol\,kg^{-1}}$). Furthermore, the authors show that none of the solvents mentioned above causes observable changes in the absorption spectra. As an ionic liquid provokes the risk of excited state quenching due to population of dark charge transfer states from the excited state, thereby, obstructing luminescence studies, STOR was excluded from the list of possible solvents. From the list of conventional solvents, DMAC, NMP, and THF are considered to by highly toxic [JA19]. Hence, DMSO is left as a reasonable choice with adequate solubility. As DMSO shows no photoluminescence (PL)a distinct ZnPc PL signal is expected in the optical studies. For comparison to the low concentration PL studies, a Raman spectrum for comparison of neat DMSO of the same batch used for the ZnPc solubility studies is discussed in appendix D.2.

5.2.1. UV/Vis Absorption

According to section 3.4.3, Transmission measurements have been performed simultaneously on a ZnPc-DMSO sample and a neat DMSO reference using two identical Hellma 100-QS macro cells with a transmission range from 200 nm to 2500 nm and an optical path length of 10 mm. For ZnPc concentrations[1] of $1 \cdot 10^{-7}\,\mathrm{mol\,l^{-1}}$,

---

[1] For comparison: The maximum concentration of ZnPc in DMSO at 20 °C is $c = 6.14 \cdot 10^{-4}\,\mathrm{mol\,l^{-1}}$ [GKR12]

$1 \cdot 10^{-6}$ moll^{-1}, and $5 \cdot 10^{-6}$ moll^{-1} a full transmission spectrum could be recorded. At lower concentrations, the absorption was found to be not sufficient while for higher concentrations the solution became opaque at certain wavelengths, leading to signal cut-offs. For each wavelength a linear fit was used to determine the molecular extinction coefficient according to equation (3.16).

In figure 5.1 a) the extracted molar extinction coefficient is plotted as a function of photon energy. The standard error is obtained by the least square fitting routine. Two main transition regions can be identified at around 1.8 eV (Q band) and at 3.6 eV (B band), well in accordance with previous studies in DMSO [ZX93][Sav+08][Cal+19].

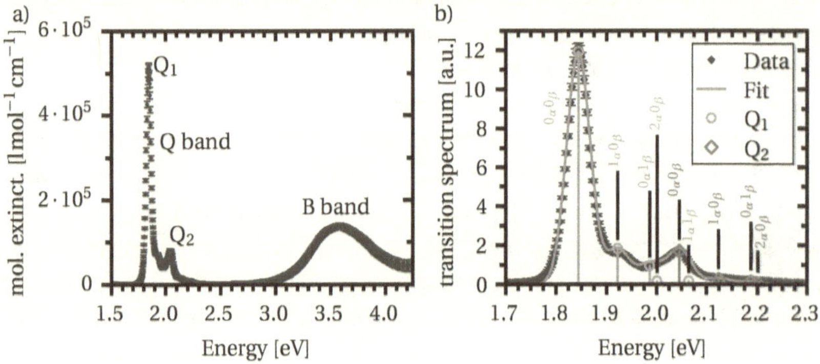

Figure 5.1.: a) Molar exctinction coefficient of ZnPc solved in DMSO. Q$_1$ and Q$_2$ indicate the two transitions related to the Q band. b) Transition spectrum of the Q band absorption and the Franck-Condon fit curve. Stick spectra labeled by the final vibrational state show the individual vibronic transitions for Q$_1$ and Q$_2$.

The Q band comprises two distinguished main transitions whose origin is still under debate and will be discussed in detail below. The B band reaching well into the UV range is not of central interest for this work. The transition is mostly ascribed to an $S_0 \rightarrow S_2$ transition [FCG16] with its main contribution being a $\pi \rightarrow \pi^*$ transition from the lower lying Gouterman orbitals (below the ZnPc's HOMO) to the LUMO [SGD73][MS95]. More recently, an additional $n \rightarrow \pi^*$ transition from the ZnPc's aza system to the LUMO is suggested to contribute to the B band [RRB01][Nem+07] as well as a $\sigma \rightarrow \pi^*$ transition [WWC17].

The Q band absorption is usually divided into two main transitions Q$_1$ and Q$_2$ as shown in figure 5.1 a). The energetically lower transition is unambiguously assigned to a HOMO-LUMO $\pi \rightarrow \pi^*$ transition [SGD73][RRB01][The+15][WWC17]. Some authors assign Q$_2$ to a vibronic transition of the Q$_1$ main transition [Sav+08]. While for a long time, the Q$_2$ transition was ascribed to a symmetrically forbidden but vibrationally allowed $n \to \pi^*$ transition [SGD73][MS95], most authors reporting on calculations of the ZnPc's electronic transitions focus their calculations on the Q_1 transition while simultaneously stressing, even though a physico-chemical conclusive explanation for the Q_2 transition is still lacking, that they cannot find a $n \to \pi^*$ band in the given energy range related to a possible Q_2 transition [RRB01][Nem+07][WWC17]. Theisen *et al.* suggested a symmetry breaking for the HOMO-LUMO transition lifting the degeneracy of the two symmetrically equivalent LUMO orbitals leading to two distinct $\pi \to \pi^*$ transitions with energies matching the Q_1 and Q_2 transitions. A detailed zoom-in of the Q band transition spectrum normalized to an area equal to one and a fit with a Franck-Condon model discussed below is shown in figure 5.1 b) offering further details, in particular, the related vibronic transitions.

In this work the, Theisen interpretation of the ZnPc absorption spectra is favored as will be discussed in the following. First of all, if the Theisen two transition model is incorporated into a fitting routine with two vibronic modes described in detail below, the Q band absorption is very well reproduced as shown in figure 5.1 b). To resemble the transition spectrum two electronic transitions Q_1 and Q_2 were assumed, each accompanied by a vibronic progression based on the displaced harmonic oscillator model. Within this model, the intensity distribution is governed by a Poisson distribution based on the Huang-Rhys parameter given by equation (2.40). However, assuming just a single dominant vibronic mode could not reproduce the full absorption spectra. Therefore, an additional vibronic mode has to be added to the modeling function. The analytical function of the transition spectrum for fitting the Q band absorption reads

$$\bar{I}(E) = A \sum_{m=\alpha,\beta}\sum_{i_m=0}^{5} \Gamma\left(E; E_{0,Q_1} + i_\alpha \cdot E_{\text{vib},\alpha} + i_\beta \cdot E_{\text{vib},\beta}, \sigma\right) \prod_{k=\alpha,\beta} \frac{e^{-S_k}}{i_k!} S_k^{i_k}$$

$$+ (1-A) \sum_{m=\alpha,\beta}\sum_{i_m=0}^{5} \Gamma\left(E; E_{0,Q_2} + i_\alpha \cdot E_{\text{vib},\alpha} + i_\beta \cdot E_{\text{vib},\beta}, \sigma\right) \prod_{k=\alpha,\beta} \frac{e^{-S_k}}{i_k!} S_k^{i_k} \qquad (5.1)$$

where Γ represents the Gaussian line shape function

$$\Gamma(E; E_\text{c}, \sigma) = \frac{1}{\sqrt{2\pi\sigma^2}} \exp\left(-\frac{(E-E_\text{c})^2}{2\sigma^2}\right). \qquad (5.2)$$

Equation (5.1) is based on the Franck-Condon description of a multiple vibronic transition starting from an initial vibrational ground state as given by equation (2.41). Within the displaced harmonic oscillator approximation, the Franck-Condon factor is given by (2.40). A total of five vibronic transitions for each vibrational mode was considered, where the cut-off vibrational number was chosen arbitrarily for very small intensities. Equation (5.1) consists of two parts, led by the factors A and $(1-A)$ representing the amplitude of the Q_1 and Q_2 transition, respectively. The parameter A determines the intensity distribution between the two electronic transitions, as the total area under the transition spectrum has been normalized to one. The following two sums in each term permute over the quantum numbers i of the two vibrational modes termed α and β hence, creating every possible (i_α, i_β)-combination and thus, determining the position of the Gaussians' central energy by shifting the 0-0 transition lines at E_{0,Q_1} and E_{0,Q_2} by the respective vibrational energies $E_{\text{vib},\alpha}$ and $E_{\text{vib},\beta}$. The last factor in each term of equation (5.1), contains the product of the Franck-Condon factors given by the Poisson distribution with the respective Huang-Rhys parameters S_α and S_β. The transition lines are assumend to be identically broadened by an energetic disorder parameter σ.

A total of eight parameters are involved in fitting the resulting function: The two main transition energies (E_{0,Q_1}, E_{0,Q_2}), two vibrational energies $(E_{\text{vib},\alpha}, E_{\text{vib},\beta})$ with their related Huang-Rhys parameters (S_α, S_β), the intensity distribution A and the energetic disorder parameter σ. As different parts of the spectra determine different parameters, no interference is observed in the fitting routine, indicated by the very small entries in the co-variance matrix, ranging from 10^{-8} to 10^{-5}. The obtained parameters are summarized up in table 5.1.

Transition	Q_1	Q_2
Energy $E_{0,q}$ [eV]	1.844	2.045 ± 0.001

Vibronic mode	α	β
E_{vib} [meV]	78 ± 1	142 ± 2
S	0.151 ± 0.003	0.078 ± 0.004

Fit parameters	$A = 0.877 \pm 0.003$	$\sigma = (23.1 \pm 0.1)\,\text{meV}$

Table 5.1.: Parameters obtained by the two mode Franck-Condon fit given by equation (5.1) and applied to the experimental ZnPc absorption spectrum.

The energies of the two main transitions are in close agreement with the theoretical values of Theisen *et al.* [The+15] (1.88 eV and 2.10 eV) and the experimental data determined for ZnPc in THF in the same study (1.86 eV and 2.07 eV). The vibrational energies of the two vibronic transitions correspond well with the pronounced ZnPc vibrations at $648\,\text{cm}^{-1}$ (80.3 meV), associated with a benzene deformation mode combined with a N,C pyrrole vibration [Pal+95], and at $\approx 1140\,\text{cm}^{-1}$ (141.3 meV), corresponding to a breathing-stretching mode of the pyrrole and benzene groups [Pal+95][Jen+84]. The HOMO is spread over the isoindole parts of the molecule an subject to the four fold rotational symmetry, while each of the degenerated LUMOs is distributed only across the opposing isoindole groups as well as the inner ring [The+15][WWC17]. Hence, the involvement of the above mentioned vibrations in the electronic transitions is highly plausible. In figure 5.1 b), the stick spectra of the vibrational transitions of the two Q transitions are depicted color coded. The label indicate the respective quantum number of the final state indexed by the respective vibrational mode. Taking a closer look at the Q_1 spectra, the need to include the β-mode to fully describe the spectra becomes obvious. As the energy as well as the Huang-Rhys Parameter of the α-mode are determined by the intensity of the $1_\alpha 0_\beta$ transition, the intensity of the subsequent $2_\alpha 0_\beta$ band determined by the Poisson distribution is not sufficient to explain the absorption transition spectrum between 1.95 eV to 2.00 eV. The spectrum is, however, very well reproduced by a superposition of the $0_\alpha 1_\beta$ and $2_\alpha 0_\beta$ absorption bands. The combined $1_\alpha 1_\beta$ transition lies fully within the Q_2 band part and contributes little to the overall shape of the absorption spectrum.

The good agreement of the applied model with the data is, nevertheless, just one reason to favor Theisen's model over other possible explanations. It could be argued that instead of introducing a second electronic transition Q_2 a third vibrational mode would be a suitable alternative explanation. Indeed, including an additional mode γ with a vibrational energy of 200 meV and a Huang-Rhys parameter of 0.145 does reproduce the experimental spectrum as well. The respective fitting parameters and the resulting curve is provided in appendix D.1. Taking into account experimental data presented later in this work, *i.e.* the emission spectra of ZnPc solutions violating the mirror symmetry law (*c.f* section 5.2.2) and the absorption of crystalline ZnPc aggregates showing two distinct transitions which cannot be explained by a Davydov splitting (*c.f.* section 5.4.1), a second molecular electronic transition gives a coherent single explanation for all the observed phenomena without the need to employ additional phenomena for each observation. Nonetheless, a third vibronic component may still be present but its transition strength is too low to be dominant in the measured absorption spectra. As mentioned before, a suggested $n \rightarrow \pi^*$ has not been reproduced in many calculations on the ZnPc absorption. Furthermore, the absorption measurements in formic acid performed by Ghani *et al.* [GKR12] showed a clear signal of a Q_2 absorption band. Acidic environments are known to dramatically blue shift $n \rightarrow \pi^*$ transitions due to protonation of the non-bonding electron pair of the nitrogen atom [Kas50] which was not observed in the respective study.

The results presented in this section, together with the current state of knowledge in literature and experimental data presented later in this work, lead to the conclusion that a lifted LUMO degeneracy leads to two distinct $\pi \rightarrow \pi^*$ transitions which explains the two Q components observed in the ZnPc's spectral absorption.

5.2.2. Excited State Emission

Photoluminescence studies have been performed according to section 3.4.2 with a Hellma 101-QS macro cell and a Thorlabs 635 nm cw-laser diode for excitation. The excitation occurs into the vibronic 1-0 band of the Q_1 absorption, about 80 meV below the Q_2 transition. As photon emission, according to Kasha's rule, originates from the radiative transition of the lowest excited state and as the laser excitation, to a large extent, only excites the Q_1 transition, a contribution from a radiative Q_2 emission to the recorded spectra can be excluded.

Figure 5.2 a) depicts the PL transition spectra of ZnPc in DMSO at a concentration of $1 \cdot 10^{-8}$ moll^{-1}. A strong main transition line at around 1.83 eV is accompanied by vibronic progressions. At around 1.58 eV two sharp Raman lines at a relative, *i.e.* with respect to the excitation wavelength, wavenumber of 2961 cm^{-1} and 3043 cm^{-1} superimpose the smooth PL intensity distribution. Their energetic splitting and relative intensities fit well to the symmetric and asymmetric C–H methyl stretching modes of DMSO [HC61]. Compared to the neat DMSO Raman spectrum (*c.f.* appendix D.2) the lines are blueshifted by approximately 50 cm^{-1}. As DMSO is highly hygroscopic [Ell+05] and the water content in the sample is known to influence the DMSO's vibrational modes [SGK73], slight deviations between different samples are expected.

A comparison between the $1 \cdot 10^{-8}$ moll^{-1} emission transition spectrum and the absorption transition spectrum obtained above (*c.f.* section 5.2.1), both normalized to the maximum transition, is shown in figure 5.2 b). As can be seen, the mirror symmetry law between the absorption and emission transition spectra is violated. The main transition lines are separated by a small Stokes shift of approximately 20 meV. The subsequent shoulders corresponding to the first vibronic transition have a similar energetic offset, but differ in intensity. The most remarkable difference can be detected for the second transition maximum at 1.65 eV (emission) and 2.05 eV (absorption). For the absorption, the peak originates from the alleged Q_2 HOMO-LUMO transition about 200 meV above the Q_1 main transition. For the emission, the relevant transition is only shifted by 170 meV with respect to the main transition. As the Q_2 electronic transition is not likely to be the cause of this emission band, it is ascribed to a vibronic transition seemingly not expressed to the same extent in the respective spectral absorption. The higher intensity for the first vibronic emission band (1.75 eV) and the occurrence of a new vibronic band lead to the conclusion that the symmetry breaking and the associated change in the molecular geometry vary the vibronic coupling in the absorption versus that in the emission process. As stated before, it is assumed that a third high energy vibrational mode is also participating in the photon absorption process but to a much lower extent. Hence, it is not visible beneath the strong intensity of the Q_2 transition band. The measured emission spectra are consistent with previous findings in the literature. For instance, in time resolved photon counting experiments a single exponential decay with a life time of about 3 ns to 4 ns is reported, indicating a single process to be mainly responsible for the observed emission [STM04][Zha+09].

To simulate the emission spectra, a Franck-Condon displaced harmonic oscillator model is used. To account for the two vibrational modes α and β found for the photon absorption and the presumed high energy vibrational mode observed in the spectral emission, hereafter called γ, a total of three vibrational modes are considered. According to equation (2.40) the corresponding Frank-Condon emission spectra can be expressed by

$$\bar{I}(E) = \sum_m \sum_{i_m=0}^{5} \Gamma\left(E; E_0 - i_\alpha \cdot E_{\text{vib},\alpha} - i_\beta \cdot E_{\text{vib},\beta} - i_\gamma \cdot E_{\text{vib},\gamma}, \sigma\right) \prod_{k=\alpha,\beta,\gamma} \frac{e^{-S_k}}{i_k!} S_k^{i_k} \tag{5.3}$$

with $m \in \{\alpha, \beta, \gamma\}$ and Γ as a Gaussian function given by equation (5.2). Under the assumption that the vibrational energies of the α and β modes do not differ from the photon absorption case, equation (5.3) contains six free parameters: The electronic transition energy E_0, the Huang-Rhys parameters of the respective vibrations S_α, S_β, and S_γ, the vibrational energy of the introduced γ mode $E_{\text{vib},\gamma}$, and the energetic disorder parameter σ. The vibrational energies of the α and β mode are fixed to the values extracted from the absorption measurements and listed in table 5.1. Again, a total of five vibronic transitions for each vibrational mode was considered. The curve (orange line in figure 5.2 a)) obtained by fitting equation (5.3) to the $1 \cdot 10^{-8} \, \text{mol} \, \text{l}^{-1}$ emission transition spectrum reproduces the data to a large extent. The stick spectra, labeled by the corresponding final vibrational state number, show the composition of the individual transition bands and reveal that the second emission maximum mainly comprises the first vibronic transitions of the β and γ mode while the low energy peak side is dominated by superimposed combined transitions, *i.e.* the final state is described by more than one vibrational mode. The resulting fit parameters are listed in table 5.2. The main transition energy reveals an small Stokes shift of 18 meV with respect to the Q_1 absorption implying an only small geometric relaxation in the first excited state. The energetic disorder shows a similar value, indicating that the energetic broadening is caused by the same underlying mechanism. The additional γ mode has a vibrational energy of (180 ± 1) meV placing it 20 meV below the Q_2 transition in the absorption spectra. The vibrational energy is in good agreement with a $1443 \, \text{cm}^{-1}$ vibration associ-

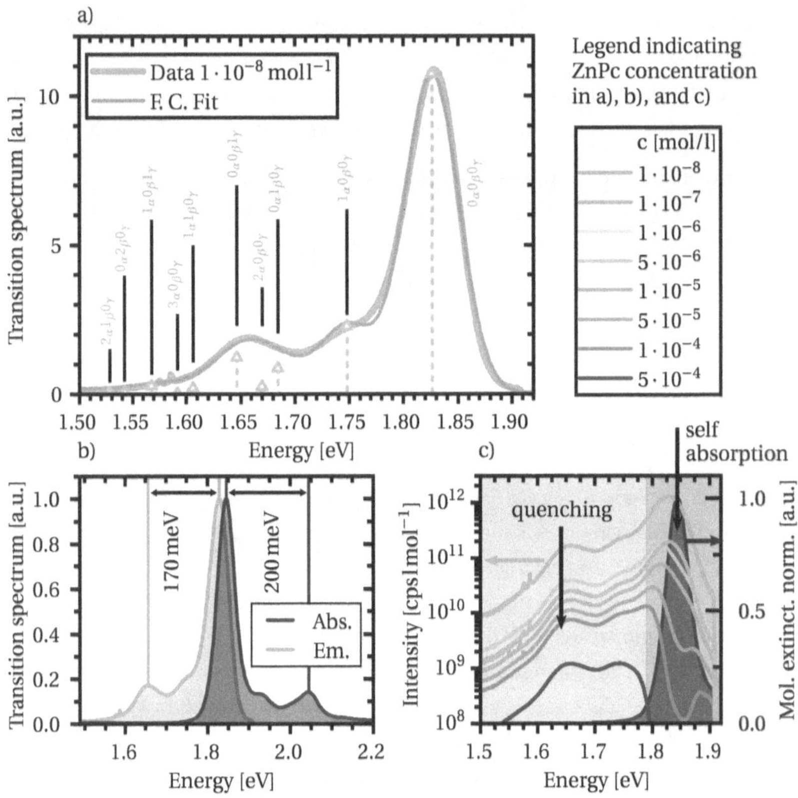

Figure 5.2.: a) Emission transition spectra normalized to an area of one for a ZnPc concentration of $1\cdot10^{-8}\,\mathrm{moll^{-1}}$. The fitted Frank-Condon curve resulting from equation (5.3) shows a good agreement with the displayed data. Stick spectra indicate vibronic transitions and are labeled by the final vibrational state. b) Emission and absorption transition spectra revealing a violation of the mirror symmetry law. c) Concentration dependent photoluminescence normalized to the respective concentration (left y-axis, blue to green color gradient) and molecular extinction coefficient (right y-axis, purple curve). Main processes governing the spectral intensity decrease in different parts of the spectra are self quenching on the low energy side (orange background) and self absorption at the high energy side (purple background).

ated with an isoindole deformation mode (pyrrole deformation and benzene twisting) [Pal+95] again in good agreement with the LUMO distribution. Comparing the vibrational energy of the γ mode with the one obtained from the three mode absorption fit (*c.f.* appendix D.1) yields a deviation in energy of the two modes. This corroborates the assumption that the Q_2 absorption is caused by a second electronic transition rather than by a vibronic transition as found for the PL spectra.

E_0	σ	S_α	S_β	$E_{vib,\gamma}$	S_γ
1.826 eV	24.8 meV	0.211 ± 0.001	0.081 ± 0.002	(180 ± 1) meV	0.114 ± 0.001

Table 5.2.: Fit parameters obtained by the three mode Franck-Condon model described by equation (5.3).

Comparing the Huang-Rhys parameters to those of the absorption, S_α increases by 40 %, while S_β does not change within the error. For the γ mode the increase in the Huang-Rhys parameter cannot be quantified, but is assumed to be significant as its contribution to the absorption process is neglectable. Although, as indicated by the small Stokes shift, the excited state shows no pronounced intra molecular relaxation, the change in vibronic coupling along certain vibrational coordinates is quite significant implying a significant bond softening. By virtue of the small Stokes shift, the LUMO of the neutral ground state can be used as an approximation for the wave function of the excited electron. This indicates that it is most likely located along two opposed isoindole groups, leaving the remaining two isoindole groups electron deprived. This leads to a bond weakening in these parts of the molecule compared to the HOMO, which is located symmetrically on all four isoindole parts of the molecule [The+15][WWC17]. The comparison with the absorption data indicates an additional relaxation process after photo excitation, presumably increasing the bond softening on the electron deducted parts of the molecule prior to photon emission.

In figure 5.2 c) the spectrally resolved PL intensity normalized to the ZnPc concentration (left y-axis) is shown for concentrations ranging from $1 \cdot 10^{-8}$ moll^{-1} to $5 \cdot 10^{-4}$ moll^{-1} as well as the normalized Q_1 molar extinction coefficient (right y-axis). Two main effects can be observed. Above 1.8 eV the spectra changes dramatically with increasing concentration while below 1.8 eV, although the overall intensity decreases, the positions of the spectral features remain fixed. Comparing the emission spectra with the Q_1 absorption reveals the cause of these changes in the spectra with increasing concentration. As more molecules are present the small Stokes shift and the resulting high overlap of absorption and emission leads to a pronounced self absorption of the emitted photons by other ZnPc molecules. For a concentration of $5 \cdot 10^{-5}$ moll^{-1} and $1 \cdot 10^{-4}$ moll^{-1} a distinct dip at 1.85 eV matching the maximum of the extinction coefficient occurs. Above these concentrations the main transition is fully suppressed. The spectral features below 1.8 eV originate from vibronic transitions and are not affected by the self absorption process. Hence, no shifts in energetic positions are observed. The overall decrease in intensity is ascribed to a self quenching process caused by the increasing probability for collisions between ZnPc molecules which lead to non radiative depopulation of the excited states, *e.g.* via transfer of excited state energy to high vibrational modes of the collision partners' ground state.

5.3. Solid State Crystal Structure

Before discussing the photophysics of the ZnPc α and β polymorph and its dependence on the molecular packing, the crystal structure of the single crystals is analyzed via XRD[2].

The impact of the $\alpha \to \beta$ phase transition on the crystal structure, crystallinity, and thin film morphology is analyzed by combined AFM and XRD studies.

5.3.1. Single Crystal Growth and Structure

ZnPc single crystals were grown via horizontal vapor deposition (*c.f.* section 3.2.1). About 20 mg to 50 mg of two fold gradient sublimed ZnPc was placed either in an aluminium or a silica glass combustion boat. The material was sublimed at (475 ± 10) °C in a 30 sccm inert N_2 gas flow. The inner tube of the furnace made of DURAN® can be heated up to 500 °C [DWK17]. As the sublimation rate is very low for the chosen temperature, the growth has to be conducted over a period of about two to four days to achieve a yield of crystals of sufficient size. Several metal phthalocyanines are known to have strongly anistropic thermal expansion coefficients [UWD43], which is why the cooling down to room temperature was accomplished slowly over several hours.

Figure 5.3 shows a representative results of the growth process. In a) a part of the inner growth tube is depicted. A bundle of thin, needle like crystals extending over 3 cm grown from the supersaturated ZnPc vapor at the onset of the declining temperature gradient (growth zone) can be seen. Towards the cooler part of the tube, the crystal growth develops into a uniform deposition of material on the tube wall (deposition zone). Usually, the formation of such a film at the colder end of the growth zone either hints to volatile impurities in the starting material or the penetration of supersaturated vapor into a temperature zone far from crystal growth conditions. However,

---

[2]Crystal growth and XRD measurements have been performed by Larissa Lazarov, Julius-Maximilian University Würzburg, as part of her final book under the author's supervision.

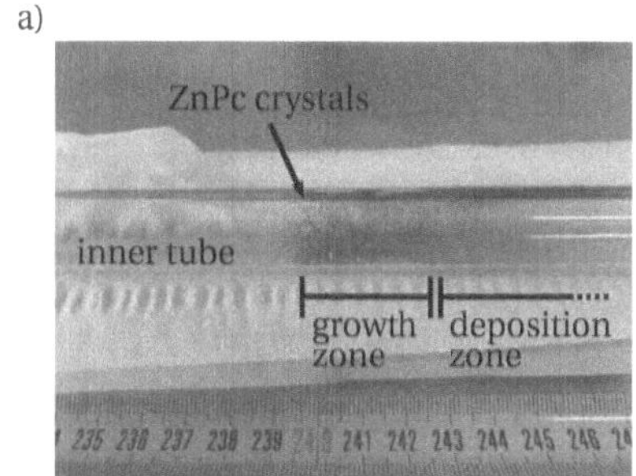

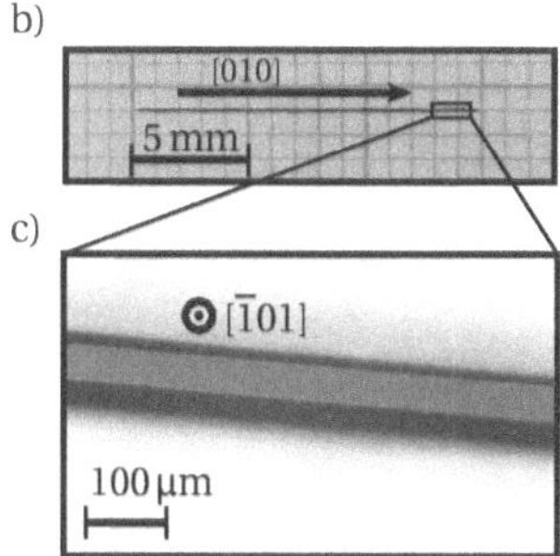

Figure 5.3.: a) Photograph of the inner growth tube with the crystal growth zone and the subsequent deposition zone. Scale of the ruler shown in cm. Material transport in furnace proceeded from left to right. b) As-grown needle like ZnPc crystal on graph paper. c) Optical microscope image of high quality smooth ($\bar{1}$01) crystal surface (20× magnification). Credits b), c): Larissa Lazarov.

contamination of the starting material is unlikely due to the high persistence of phthalocyanines against atmospheric oxidation [MT63] and the fact that the observed clear separation along the temperature gradient would have made potential contamination easily removable by the twofold gradient sublimation purification performed.

Figure 5.3 b) shows a photograph of a ZnPc crystal on graph paper. According to Luc *et al.* [Luc+20] ZnPc single crystals grow in the stable β phase and the crystal's long axis matches the crystallographic [010] direction. This crystallographic direction corresponds to the π-stacking direction (*c.f.* figure 3.2) leading to an energetically favored fast growth of the molecular stacks along the [010] axis. These results could be confirmed by independent XRD studies on the sublimation grown crystals performed by Dr. Krzysztof Radacki, Insitute of Inorganic Chemistry, Julius Maximilians University Würzburg. The determined crystal structure is listed in table 5.3 and matches the published data of [Luc+20] in table 3.1. Usually, the needles have a rectangular cross section, with the wider facet defining the ($\bar{1}$01) surface. The optical microscope image (20× magnification) in figure 5.3 c) shows a representative ($\bar{1}$01) facet with a smooth and visually defect free surface over several hundred of μm

The crystallographic lattice plane corresponding to crystal surface shown in figure 5.3 c) was determined to ($\bar{1}$01) by measuring its out-of-plane Bragg reflection. The XRD pattern is shown in figure 5.4 a). The diffraction peak correspondence was confirmed by comparing the obtained data with diffraction patterns simulated by the free software package PowderCell [KN96] based on the crystallographic data in [Luc+20] and table 5.3. The corresponding molecular orientation is shown in figure 5.4 b). The ZnPc

Cell lengths		Cell angles		Parameters	
a	14.5297(3) Å	α	90°	S. g.	P $2_1/n$ (14)
b	4.84010(10) Å	β	106.301(2)°	V	(1156.12 ± 0.04) Å^3
c	17.1282(4) Å	γ	90°	Z	2

Table 5.3.: Unit cell parameters determined by Dr. Krzysztof Radacki, Julius Maximilians University Würzburg: cell lengths (a, b, and c), cell angles (α, β, and γ), symmetry group (S.g.), cell volume V, and molecules per unit cell Z.

molecules are arranged in a herringbone like pattern with the individual stacks oriented along the crystallographic [010] direction.

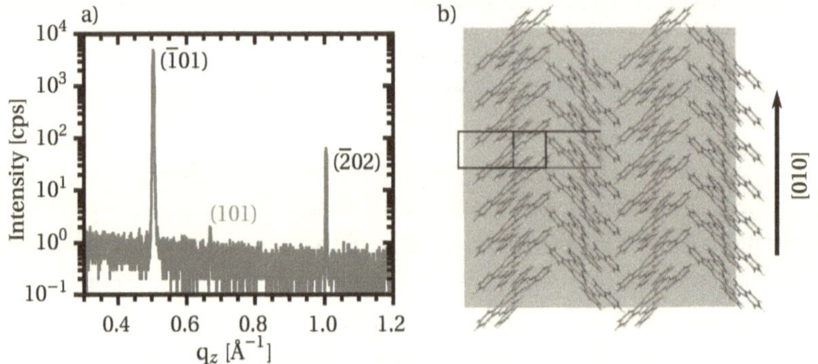

Figure 5.4.: a) XRD pattern of the ($\bar{1}$01) surface of a sublimation grown ZnPc crystal in the β phase similar to the one shown in figure 5.3 c). b) Molecular orientation in the ($\bar{1}$01) plane visualized with the Mercury 4.0 software package [Mac+20].

The X-ray diffractogram in figure 5.4 a) shows an additional weak diffraction peak at $q_z = 0.669$ Å^{-1} which has been identified as a (101) reflection. A comparison with simulated XRD powder diffraction pattern by Mercury 4.0 software package [Mac+20] shows that the intensity ratio to the ($\bar{1}$01) main peak is approximately $5 \cdot 10^{-4}$ times weaker than expected for a generic β phase ZnPc XRD powder pattern. However, the intensity of the consecutive ($\bar{2}$02) diffraction in comparison to the ($\bar{1}$01) diffraction peak agrees with the intensity distribution expected from the powder pattern simulation. Therefore, the residual (101) reflection is attributed to a crystal defect.

5.3.2. Thin Film Structure and Phase Transition

The morphological and structural characterization presented in this section have previously been published in [Ham+19]. As, at the time of the publication, the ZnPc β phase crystal structure had not been published the analysis is conducted by comparison to the copper phthalocyanine's (CuPc) crystal structure published by C. J. Brown [Bro68]. Both structures are an equivalent representation of the identical unit cell as they have the same reduced cell parameters, which has been verified with the CELL-TRANS program from the Bilbao Crystallographic Server [Aro+06b][Aro+06a][Aro+11]. However, the unit cell reported by Brown associates the (001) reflection with the out-of-plane Bragg reflection of the above analyzed ZnPc crystal surface, whereas the unit cell published by Luc *et al.* yields a ($\bar{1}$01) correspondence for the same peak. For consistency with the already published data, the (001) notation is retained for the discussion of the thin film structure, however, emphasizing that the out-of-plane direction of the thin film and that of the ZnPc single crystal surface analyzed in this work correspond to the same crystallographic direction.

Silicon wafers covered by a thermally grown SiO_2 layer were used as substrates for ZnPc thin film growth. The films were prepared via high vacuum sublimation of twofold gradient sublimation purified ZnPc at a deposition rate of $0.17\,\text{Å}\,\text{s}^{-1}$ up to a nominal film thickness of 30 nm. The $\alpha \rightarrow \beta$ phase transition has been thermally induced in an annealing furnace as described in section 3.4.2. The samples were kept at 250 °C under inert gas flow while simultaneously measuring PL spectra over the course of two hours until the recorded spectra indicated the structural phase transformation to be completed. A detailed discussion of the PL data is given in section 5.5.1. As discussed below, the as deposited samples adapt the crystallographic α phase, while the thermal annealing leads to a transformation to the β phase as previously reported in [Gon+18][AC16].

In figure 5.5 a) the small angle X-ray reflection pattern is shown for an as deposited thin film sample (α phase) and the thermally annealed sample (β phase). Above the critical angle (here, at momentum transfer $q \approx 0.03\,\text{Å}^{-1}$), three Kiessig fringes are visible for both samples. After correction for the Fresnel reflectivity (equation (3.9)), the distance between the pairs of adjacent maxima or minima, is used to calculate the film thickness according to equation (3.10), yielding film thickness d of $(290 \pm 10)\,\text{Å}$ and $(360 \pm 20)\,\text{Å}$ for the α phase and β phase films, respectively, as listed in table 5.4. The film thickness of the as deposited sample is in excellent agreement with the nominal thickness measured by the quartz crystal microbalance during deposition. For the β phase, *i.e.* after thermal annealing, the film thickness is slightly enhanced.

The first order Bragg reflection of both samples is depicted in figure 5.5 b). Both samples show a clear main maximum, accompanied by subsidiary maxima. These Laue oscillations indicate a high out-of-plane crystallinity of the thin film domains. The Laue function (3.8) together with a Fresnel prefactor (3.9) was fitted to the first order Bragg

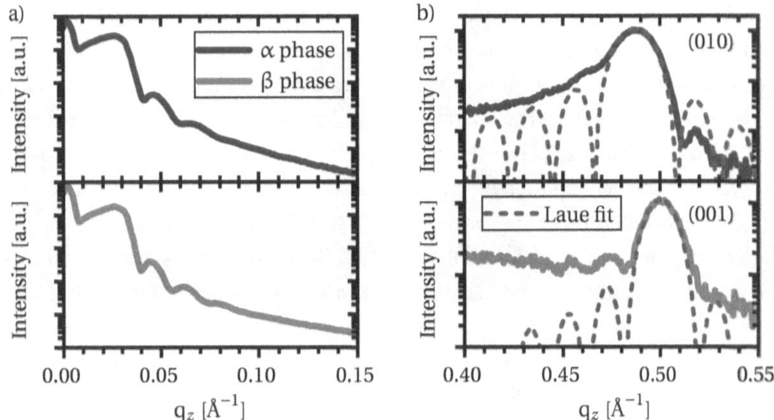

Figure 5.5.: ZnPc thin films in the as deposited α phase (purple curves) and β phase (red curves) after annealing: a) XRR data of both thin film samples. Above a critical momentum transfer at $0.03\,\text{Å}^{-1}$ Kiessig fringes are visible for both types of samples. b) First order Bragg reflection with Laue oscillations indicating a high crystallinity of both samples. Laue fit functions are indicated as dashed blue curves. Data has been previously published alongside [Ham+19] as supplementary information.

reflection of both samples with an estimated variation in the number of lattice planes of ±1. With the lattice distance obtained by the Laue fit the average crystallite heights can be estimated by $N \cdot d_z$ to $h_\alpha = (297 \pm 13)\,\text{Å}$ and $h_\beta = (338 \pm 13)\,\text{Å}$ for the α phase and β phase thin film samples, respectively, and are listed in table 5.4 together with the respective d_z values. The respective crystallite heights match the film thicknesses obtained by the evaluation of the Kiessig fringes, indicating that the individual crystallites extend over the whole film thickness. For the annealed sample the increase in the film thickness and mean crystallite height is ascribed to a thermally induced dewetting of the ZnPc thin film, *i.e.* molecular diffusion and agglomeration during the annealing process [Kra+03]. Finally, the obtained lattice spacings enable a first distinction and assignment of the two distinct structural phases. For the as deposited sample, an assignment of the ZnPc α phase is possible, as the lattice distance of $d_z = 12.9\,\text{Å}$ corresponds to the (010) lattice distance of $d_{010} = 13\,\text{Å}$ [Erk04][Bre+15]. For the annealed sample, the lattice distance of $d_z = 12.6\,\text{Å}$ is in excellent agreement with the (001) lattice distance $d_{001} = 12.5\,\text{Å}$ of the β phase [Bro68][Luc+20].

For comparison, the first order Bragg reflections are depicted in figure 5.6 a) on a linear scale with their intensity normalized to the respective peak maximum, so that the offset between the two reflection at $q_z = 0.487\,\text{Å}^{-1}$ and $q_z = 0.500\,\text{Å}^{-1}$ becomes ob-

		Laue Function			Bragg Reflection[*]	
Struc. phase	Film thickness	Mean crystal height	N	d_z [Å]	q_z [Å^{-1}]	d_z [Å]
α	(290 ± 10) Å	(297 ± 13) Å	23 ± 1	12.9	0.487	12.90
β	(360 ± 20) Å	(338 ± 13) Å	26 ± 1	12.6	0.500	12.57

Table 5.4.: Structural data of α phase and β phase thin film samples. Film thickness obtained by XRR analysis. From the fitting of the Laue function the number of lattice planes N and lattice distance d_z can be determined, yielding the mean crystallite height as $N \cdot d_z$. The out-of-plane lattice distance d_z has been determined from first order Bragg reflection, as well. [*] q_z and d_z values published in [Ham+19] have been rounded to two and one decimal places, respectively.

vious. The corresponding lattice distances (*c.f.* table 5.4) are in good agreement with the values obtained by the Laue function and confirm the thermally induced $\alpha \rightarrow \beta$ phase transition [AC16]. Figures 5.6 c) and d) illustrate the molecular packing for both crystallographic phases. The α phase comprises parallel π-stacked molecules along the short crystallographic a axis (marked in red). While in the β phase the molecules are still π-stacked along the unit cell direction b (marked in red), the stacks pack in a mirror symmetry fashion leading to an overall in-plane herringbone like pattern [Bru+10][Gon+18]. There are several more meta stable polymorphs reported for metal phthalocyanines which differ in their molecular packing in the unit cell as well as their space group [Erk04]. However, the inconsistent nomenclature in literature obstructs the comparison with the obtained results. A particular difficulty is imposed by the inconsistent assignment of the α and γ polymorph as the respective crystal structures differ in space group, number of molecules per unit cell, and molecular packing. The terminology used here refers to that of Erk *et al.* [Erk04] with the α phase labeling a triclinic unit cell comprising one molecule and a parallel π-stack molecular packing (*c.f.* table 3.1 and figure 3.2 b)). The γ phase, sometimes referred to as α phase [Lou+07][He+15], has a monoclinic unit cell with four molecules packed in a herringbone pattern. As the number of molecules per unit cell as well as the packing strongly influence the photophysical properties of the crystalline aggregates, caution is advised when comparing results to literature data.

To compare the average angular tilting of the individual crystallites and to qualitatively probe changes in defect density of the crystalline films upon the structural phase transformation rocking scans have been performed on the first order Bragg reflection of both samples. The rocking curves normalized to their maximum are shown in figure 5.6 b). Both samples yield narrow peaks with FWHM of $\Delta_\alpha = 0.052°$ and $\Delta_\beta = 0.096°$

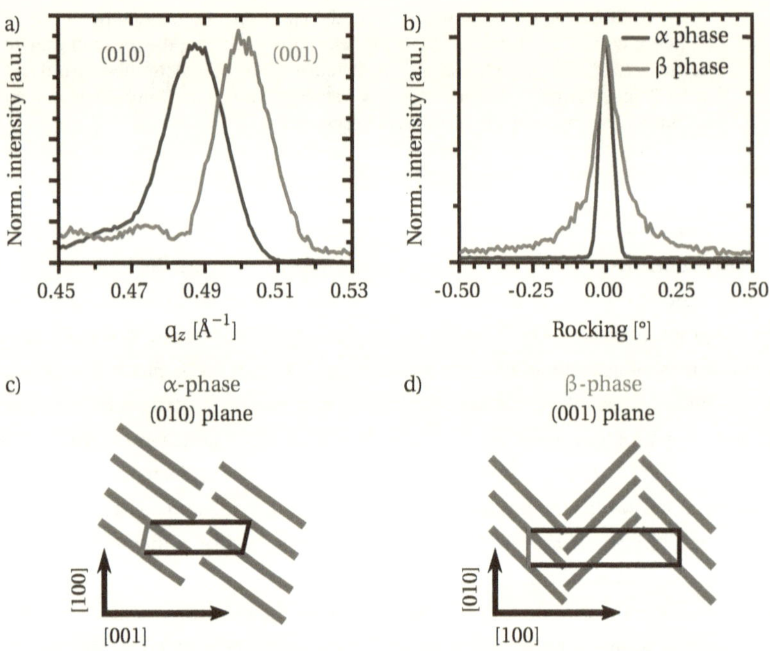

Figure 5.6.: a) First order Bragg reflection of both thin films normalized to the maximum peak intensity. The offset between the reflection peaks is clearly visible. b) Rocking curves on the first order Bragg reflection of each film suggesting a higher defect density after the $\alpha \rightarrow \beta$ phase transformation. c) and d) schematically show the molecular packing in both crystallographic phases. The π-stacking of the molecules occurs along the short unit cell dimension (marked in red) in each structural phase. Reproduced from [Ham+19], with the permission of AIP Publishing. Data in b) has been published on a logarithmic scale.

indicating that the distinct preferential out-of-plane alignment of the crystalline domains is kept upon phase transformation.

The enhanced off specular background found in the β rocking curve indicates a higher defect density in the annealed film [Nic+04]. This is corroborated by the larger FWHM of the associated rocking curve as equation (3.7) lists defect induced broadening, *i.e.* defects and strain induced around screw dislocations, as a main contribution to the rocking width. Other terms in equation (3.7) can be excluded as main contributions to the observed broadening. The instrumental broadening does not change between measurements and broadening by sample curvature can usually be neglected for substrates much thicker than the film thickness [Aye94]. Furthermore peak broadening due to crystallite dimensions is only dependent on the film thickness t by the relation $\propto t^{-1}$ and hence, the slightly larger film thickness of the β phase should, if at all, lead to a narrowing of the rocking peak [Spi+09].

To complete the structural investigations on the two crystallographic phases, AFM topography studies were performed on the thin film samples. The α phase, shown in figure 5.7 a) comprises small circular grains of 20 nm to 50 nm in diameter, forming a smooth and isotropic surface. In contrast, after transformation to the β phase, the film is composed of domains formed by oriented anisotropic needle like crystallites with lengths of 0.5 µm to 2 µm and widths of 50 nm to 100 nm. For both surface scans the height distribution is shown as insets in the upper left corner of figures 5.7 a) and b). For the smooth surface of α phase thin film displayed in 5.7 a) a symmetric, narrow distribution with a FWHM of $\Delta_\alpha = 1.9$ nm is found. Upon phase transformation the β crystallization and thermally activated dewetting lead to a broad and slightly asymmetric height distribution with a FWHM of $\Delta_\alpha = 4.1$ nm. The surface scans show that the annealing process leads to the transformation of the small α phase grains to needle like β phase crystallites of much larger extend. The interplay between the thermally activated dewetting, *i.e.* diffusion and agglomeration of molecules on the film surface, and the structural phase transformation could not be disclosed by the here presented room temperature AFM scans performed before and after annealing. Interestingly the rocking data indicates a larger density of microscopic structural defects after the annealing process, despite the fact that the AFM surface scans indicate much larger crystalline domains for the annealed β phase thin film sample.

To conclude the structural analyses, complementary studies on a ITO/glass substrates have been performed to investigate the application of ZnPc thin films of both structural phases in prototypical OLED devices as reported below (*c.f.* section 5.5.2). Figure 5.8 shows XRD measurements carried out 30 nm thick ZnPc thin film samples prepared in a similar manner but on ITO/glass substrates. Again, to induce the desired

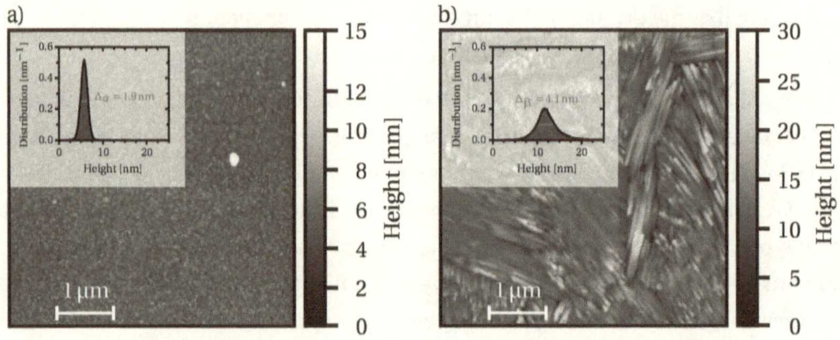

Figure 5.7.: AFM topography scans of a) ZnPc α phase thin film consisting of small circular grains and b) ZnPc β phase thin film comprising needle like crystallites. The insets in the upper left corner of each image display the height distribution of the respecitve topography scan illustrating the rougher surface for the annealed β phase thin film sample. For each distribution the FWHM Δ is indicated in red. AFM images in a) and b) reproduced from [Ham+19], with the permission of AIP Publishing.

phase transition, an as deposited sample has been heated in furnace under N_2 gas flow at 300 °C for one hour. The first order Bragg reflections are located at $q_{z,\alpha} = 0.485\,\text{Å}^{-1}$ and $q_{z,\beta} = 0.498\,\text{Å}^{-1}$ for the α and β phase sample, respectively. The resulting lattice distances of $d_z = 13.0\,\text{Å}$ for the pristine α sample and $d_z = 12.6\,\text{Å}$ for the annealed sample confirm the occurrence of the $\alpha \rightarrow \beta$ phase transition. Estimating the FWHM of both diffraction peaks on a transformed 2θ-abscissa by means of a Gaussian function taking a linear background into account, enables the estimation of the crystallite height by means of the Scherrer equation (3.6). No correction for instrumental broadening is necessary as the FWHM of the peaks amounts to approximately $5 \cdot 10^{-3}$ rad whereas an instrumental broadening of $\approx 1 \cdot 10^{-3}$ rad has been estimated [Han17] (*c.f.* 3.3.2). The crystallite heights for the α and β film sample are $h_\alpha = (230 \pm 10)\,\text{Å}$ and $h_\beta = (290 \pm 20)\,\text{Å}$, respectively. Even though the intensity is weaker compared to the films grown on SiO_2 substrates discussed before indicating an overall lower crystallinity of the films grown on ITO substrates, the out of plane crystallite height is of the same order as the nominal film thickness. Secondly, the trend of a larger crystallite heights for the ZnPc β phase films on SiO_2 is observed as well, indicating that thermally activated dewetting during the annealing is present for ZnPc films on ITO substrates as well.

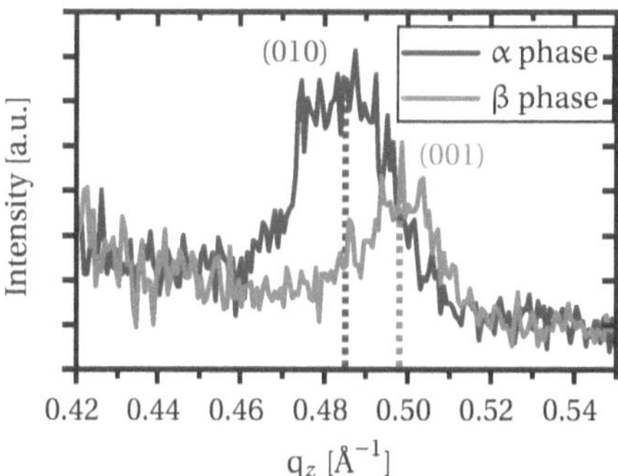

Figure 5.8.: Out-of-plane XRD of α and β phase ZnPc thin films on ITO substrates. The vertical dotted lines indicate the (010) and (001) peak position for the ZnPc α and β phase, respectively. Data has been previously published as supplementary information in [Ham+19].

5.4. Excited States in ZnPc Polymorphs

After the structural characterization of both polymorphs in thin film samples and single crystals, the excited states' photophysics are investigated via steady state absorption and temperature dependent PL studies.

5.4.1. Solid State Absorption and Emission

For measurements of the solid state absorption two 30 nm thick films have been deposited via high vacuum sublimation on ITO/glass substrates. For one sample the $\alpha \to \beta$ phase transition was induced in the gas flow furnace at a temperature of 300 °C. Steady state transmission spectra were measured and corrected for the transmission of a neat ITO/glass substrate. The absorption coefficient was obtained with equation (3.16) and the transition spectrum was calculated by scaling to the respective transition energy according to (3.12). As no correction for back reflection on the air-glass interfaces was conducted, the obtained absorption values are slightly overestimated. Nevertheless, the transition energies as well as their relative intensities can be determined reliably. Figure 5.9 a) shows the absorption transition spectrum of the ZnPc Q band for the α and β phase. Additionally, the single molecule Q_1 and Q_2 transitions in DMSO are indicated by the blue lines.

Both polymorphs show a splitting of the absorption band. Many studies have attributed the two transitions to a Davydov splitting due to inter molecular coupling [KM08], sometimes in combination with a proposed $n \to \pi^*$ transition [He+15][AC16]

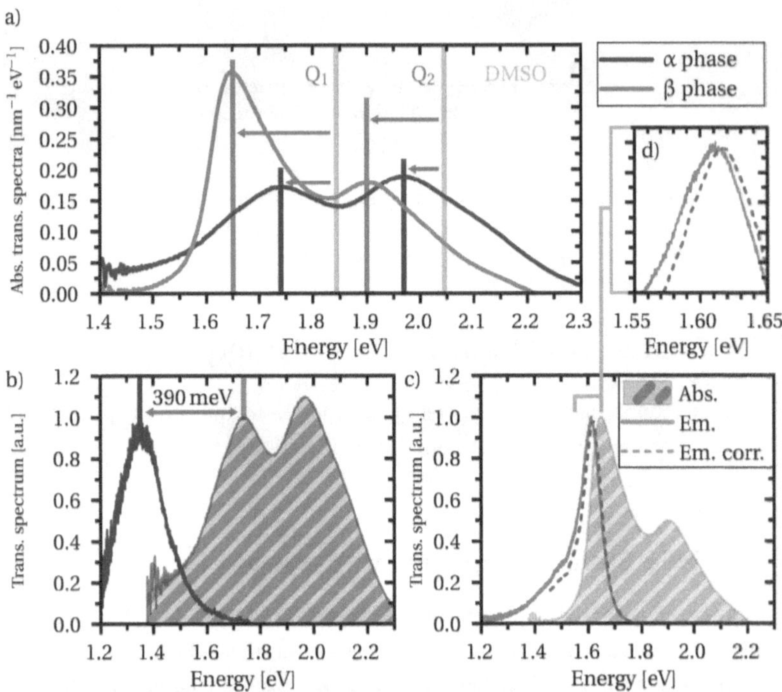

Figure 5.9.: Absorption and emission spectra of the α and β polymorphs in ZnPc thin films on ITO substrates: a) Absorption transition spectra of both polymorphs together with the single molecule Q_1 and Q_2 transitions in DMSO indicated by the blue lines. The aggregation induced red shift is indicated by arrows. b) and c) Comparison of absorption (normalized to Q_1) and emission (normalized to its maximum) transition spectra of α and β phase, respectively. For the β phase emission, a self absorption corrected (Em. corr.) spectra is shown. d) Zoom-in on the 0-0 transition of the β phase emission.

based on the Mack-Stillman assignment [MS95]. A Davydov splitting for the β and the γ phase is possible due to the transitionally nonequivalent molecules in the unit cell. However, the α unit cell is occupied by just one molecule rendering the reported Davydov splitting (*e.g.* [AC16] and [Bre+15]) impossible. The energetic splitting of 230 meV and 250 meV for the α and the β sample, respectively, are similar and close to the observed splitting of about 200 meV for the single molecules in DMSO. As the splitting is present for both polymorphs, it is likely not caused by the molecular packing but is an inherent property of the ZnPc molecule itself only modulated by inter molecular coupling upon aggregation. The two transitions are hence ascribed to the two molecular transitions Q_1 and Q_2 as described for ZnPc in DMSO (*c.f.* section 5.2.1). For comparison the transition energies of Q_1 and Q_2 as well as the respective band splittings $\Delta E_{Q1;Q2}$ are listed in table 5.5. A recently published study by Feng *et al.* [Fen+20] should not be unmentioned. Their performed calculations on an optimized crystal structure closely resembling the α phase suggest a mixing of a CT and Frenkel excitonic state leading to an excitation of hybrid character that causes the observed two peaks in the thin film absorption spectra. The varying intensity distribution observed in literature and in the present study is attributed to a change in inter molecular distance along the π-stacking direction and under this assumption, the experimental spectra are reasonably well reproduced. The second peak Q_2 in the single molecule absorption is associated with a vibronic transition as supported by calculations. However, emission spectra in relation to different solid state polymorphs have not been discussed by the authors.

For both transitions a red shift upon aggregation compared to the single molecule absorption is clearly visible. The respective shifts ΔE_{DMSO} of the two transitions Q_1 and Q_2 are 100 meV and 75 meV, respectively, for the pristine α phase ZnPc film, whereas for the annealed β phase ZnPc film the shift amounts to 194 meV for the Q_1 and 145 meV for the Q_2 transition. These results are listed in table 5.5 together with the Q_1 and Q_2 transition energies for both samples. In the β phase sample, the aggregation induced red shift of both transitions is roughly twice as large as compared to the α sample. Furthermore, the α phase transitions are similar in intensity, whereas for the β phase the Q_1 transition is dominant. The measured absorption spectra are in agreement with those obtained in similar studies [Gaf+10][AC16], even though a comparison has to be treated with care due to a missing structural evidence of the phase transition ([Gaf+10]) or an presumably erroneous assignment of the respective polymorph ([AC16]).

The PL signals of both samples at 685 nm excitation are shown in figures 5.9 b) and c). For the α phase (subfigure b)) a weak signal around 1.35 eV is detected. As the

Sample	Q_1		Q_2		$\Delta E_{Q1;Q2}$
DMSO	1.844 eV		2.045 eV		201 meV
	Q_1	ΔE_{DMSO}	Q_2	ΔE_{DMSO}	
α	1.74 eV	100 meV	1.97 eV	75 meV	230 meV
β	1.65 eV	194 meV	1.90 eV	145 meV	250 meV

Table 5.5.: Electronic transitions in ZnPc α and β polymorphs determined by thin film absorption measurements. For comparison the single molecule transitions in DMSO are listed, too. The aggregation induced red shift with respect to that of the single molecule ΔE_{DMSO} as well as the energetic splitting of the Q band components $\Delta E_{Q1;Q2}$ are given for both polymorphs.

Stokes shift is expected to be in the order of the vibrational energy [Jon+15], its resulting value of 390 meV is unusually large. In comparison, the β phase emission (subfigure c)) shows only a slight red shift and a PL spectrum otherwise comparable to the single molecule spectrum obtained in DMSO, with the difference that the second maximum in the emission spectra is missing. The absence of the second maximum in the PL further corroborates the presence of two distinct electronic transitions Q_1 and Q_2 and contradicts a potential vibronic origin of the Q_2 transition. As the spectral overlap of absorption and emission is not negligible, potential effects by self absorption PL of the β polymorph have to be considered. Irkhin *et al.* [Irk+12] reported an expression for an experimental PL spectrum $S_{rec}(E)$ as a function of energy recorded orthogonal to a surface of an "infinitely" thick crystal taking into account the deposited excitation power at the respective depth as well as self absorption of the emitted photons from the intrinsic PL spectrum $S_0(E)$. Transferring the expression to a thin film with finite film thickness t yields

$$S_{rec}(E) = S_0(E)\alpha_{exc} \int_0^t e^{-\alpha_{exc}x} e^{-\alpha(E)x} \, \mathrm{d}x$$
$$= S_0(E) \frac{\alpha_{exc}}{\alpha_{exc} + \alpha(E)} \left(1 - e^{-(\alpha_{exc}+\alpha(E))t} \right). \tag{5.4}$$

The recorded signal $S_{rec}(E)$ as a function of emission energy E is obtained from the intrinsic spectrum $S_0(E)$ by taking into account the absorption coefficient α_{exc} at the excitation energy determining the deposited excitation power and the absorption coefficient $\alpha(E)$ as as a function of the emission energy. The dashed line in figure 5.9 c) indicates the normalized emission transition spectrum corrected for self absorption in the energy range in which reliable absorption coefficients could be extracted from the data presented in figure 5.9 a). Even though the spectral overlap is significant, the recorded spectrum only slightly overestimates the intensity of the first vibronic transitions and thus, the emission maximum is only maginally blue shifted ($\approx 6\,\text{meV}$, *c.f.* zoom-in into figure 5.9 c)) yielding a corrected emission maximum at $1.617\,\text{eV}$. After correcting for self absorption, the Stokes shift is determined to $33\,\text{meV}$ which is only slightly larger compared to the single molecule spectra in DMSO.

The differences in absorption as well as in emission can be explained by the different molecular stackings in the respective unit cells. The π-stacking direction along the short unit cell axes, [100] and [010] (marked in red in figure 5.6 c) and d)), for the α and β phase, respectively, determines the photophysics of the molecular aggregates. An approximation of the underlying coupling can be deduced by assuming a point dipole-dipole interaction, even though due to the close proximity possibly other coupling mechanisms or higher multipole moments could be present as well [SB16][Sch+19]. As discussed in section 2.3.2, the slip angle β between parallel molecules determines the energy levels and transition dipole moments of the aggregates. Assuming the direction of the ZnPc transition dipole as described in [Zha+16], the slip angle between two adjacent transition dipoles was estimates for both crystal structures as the angle between a carbon atom that forms the inner C-N alternating ring (*c.f.* figure 3.1) and the central Zn atom of the same molecule and the Zn atom of the next molecule, using the angle measurement tool of Mercury 4.0 [Mac+20] with the published structures of the α phase [Erk04] and the β phase [Luc+20]. For the α unit cell, a slip angle of $\beta_\alpha = 90° - 65.38° = 24.62°$ was obtained and for the β phase, the slip angle was determined to $\beta_\beta = 90° - 45.48° = 44.52°$. The comparison with the magic angle $\beta_\text{m} \approx 35.26°$ that marks the transition from an H-aggregate to a J-aggregate (*c.f.* section 2.3.2) shows that in the ZnPc α phase the molecular stacks form an H-aggregate ($\beta_\alpha < \beta_\text{m}$), leading to a blue shift of the absorption bands with respect to the gas-to-crystal shift as well as to a suppression of the 0-0 photon emission explaining the overall red shift in the PL spectrum. As will be discussed in section 5.4.3, the residual luminescence of the α phase is caused by excimer formation which presumably becomes the dominant emission channel as the direct photon emission is dipole forbidden in the aggregate. The stacking in the β unit cell yields $\beta_\alpha > \beta_\text{m}$ and hence, implies J-aggregate formation. The dipole-dipole interaction along the [010] stacking direction leads to the observed red shift compared to the α phase and the strong Q_1 transition band. As the Q_1 absorption most likely does contains contributions of vibronic progression as indicated by the broadening of the high energy flank towards the Q_2 transition, the inter molecular coupling is not strong enough to generate a single exciton band but the vibronic states are preserved leading to the formation of vibronic bands in the excited state (*c.f.* section 2.3.3).

5.4.2. Temperature Dependent Emission

To investigate the excited state energetics and inter molecular couplings, temperature dependent PL measurements were performed[3] at an excitation wavelength of 685 nm as described in section 3.4.2. The α phase was prepared by high vacuum deposition. A 30 nm thick film was deposited on a silicon wafer covered by a native oxide layer as described before. The silicon diode for temperature control was fixed with silver paint near the edge of the wafer and the PL measurement was conducted approximately 1 cm away from the diode to minimize any influence of the silver paint's solvents on the lattice structure of the thin film. The silicon substrate was mounted with silver paint on the cryostate's cold finger for thermal coupling. A β phase single crystal and the silicon diode for temperature control were fixed with silver paint on a copper plate to ensure sufficient thermal coupling. The copper plate was mounted to the cold finger with silver paint as well. Both samples were cooled down to the lowest accessible temperature (17 K for the thin film sample, 5 K for the single crystal) and, afterwards, slowly heated up. PL spectra were recorded after defined steps at fixed temperatures. Prior to the measurements the samples were kept at the respective temperature for at least 15 min to ensure thermal equilibrium. The spectra were recorded using the 300 lines mm^{-1} diffraction grid of the spectrometer. For the thin film spectra an acquisition time of 60 s and the central wavelengths of 790 nm, 875 nm, and 960 nm were used. The single crystal spectra were taken with an acquisition time of 5 s below 140 K and 8 s above. Central wavelengths were chosen to 790 nm and 875 nm up to 80 K. Above, a spectrum centered at 960 nm was added.

In figure 5.10 the resulting transition spectra are plotted over the examined temperature range for both samples. For the ZnPc α phase presented in a), at low temperatures an asymmetric emission profile is detected with a PL maximum at 1.33 eV. With rising temperature, the FWHM of the emission increases from about 160 meV to 240 meV. This strong increase in FWHM with temperature, in combination with the overall low intensity and the observed large Stokes shift indicates excimer formation as described section 2.3.2. However, the predicted intensity decrease of the emission maximum is not observed for the ZnPc α phase. As depicted in figure 5.11 a), the integrated intensity increases with temperature, which can not be explained by sole excimer formation.

For the β phase crystal the transition spectra depicted in figure 5.10 b) shows a complex emission behavior with temperature. At low temperatures, a narrow emission peak with a FWHM of about 45 meV at 1.62 eV dominates the emission. The emission quickly decreases with rising temperature. At above 100 K, a pronounced emis-

---

[3]PL measurements on the ZnPc β phase single crystal were performed by Larissa Lazarov, Julius-Maximilians University Würzburg, as part of her final book under the author's supervision.

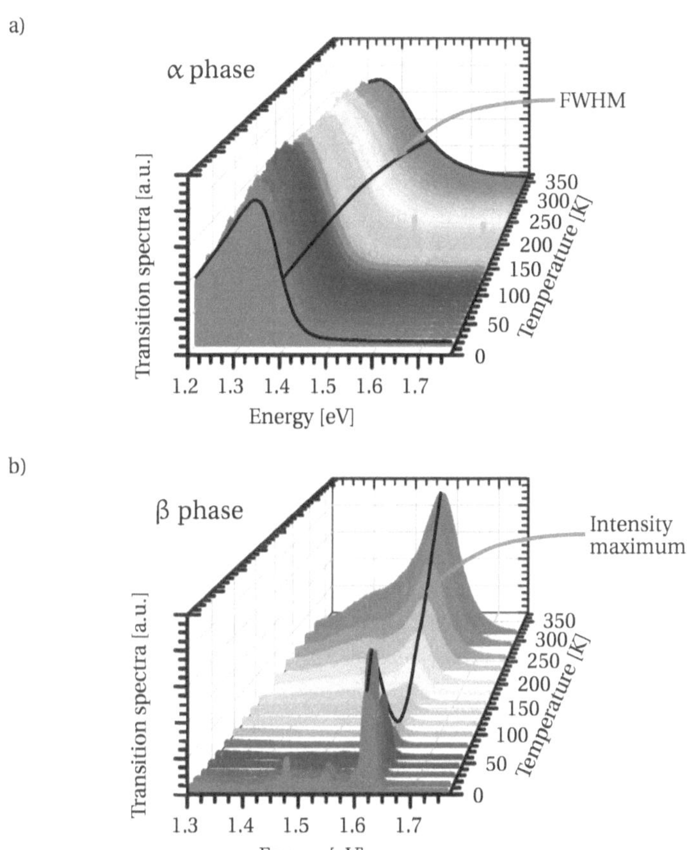

Figure 5.10.: Temperature dependent PL transition spectra of a) a ZnPc α phase thin film and b) a ZnPc β phase single crystal. Black lines indicate the spectral shape and the evolution of the FWHM (α phase) and emission maximum intensity (β phase) across the measured temperature range.

sion spectrum evolves, resembling the shape of the room temperature emission observed for β phase ZnPc layers on ITO, and increases in intensity. The main emission maximum is slightly red shifted to 1.59 eV which is most likely caused by the higher self absorption in the crystal compared to the thin film. Correcting for self absorption the alignment of the individual crystallographic directions with respect to the incident light and its polarization can be crucial (*c.f.* section 4.4.2) and has to be taken into account [Irk+12]. Hence, performing a self absorption correction using the available thin film absorption is likely prone to errors due to the rotational domain structure thus averaging of, at least, the in-plane crystallographic axis. However, as the observed self absorption effect found for thin films is small no correction is made. The emission behavior below 100 K resembles that of J-aggregates [Spa10] [Mül+13] [Eis+17] and, in particular, their superradiance (*c.f.* section 2.3.3) and will be discussed in the subsequent chapter 5.6.3.

Both samples show a counter intuitive behavior with temperature. Usually, the overall intensity decreases with increasing temperature as non radiative depopulation processes, *e.g.* phonon scattering, become activated. In figure 5.11 a) and b) the integrated intensity is plotted as a function of sample temperature. For the α phase illustrated in subfigure a) an overall increase with temperature is evident, whereas for the ZnPc β phase crystal after an initial intensity decrease with rising temperature the overall emission increases above 100 K.

For the integrated intensity of the α phase thin film below 100 K, the emission seems to be constant followed by a steady increase in intensity towards higher temperatures. A phenomenological model assuming a temperature independent radiative decay rate $k_{r,0}$ resembling the natural radiative decay rate, and a phenomenological Boltzmann activated radiative rate with activation energy $E_{\alpha,a}$ leads to the following equation for the effective radiative decay rate k_r:

$$k_r(T) = k_{r,0} + k_T e^{-\frac{E_{\alpha,a}}{k_B T}}. \tag{5.5}$$

Under the assumption that the signal is caused by a single electronic transition, the detected integrated intensity $I(T)$ is proportional to the radiative transition rate $k_r(T)$. Below 100 K the temperature independent radiative decay rate governs the $I(T)$ behavior and leads to a constant intensity $I_0 \propto k_{r,0}$. Taking I_0 as the mean value of $I(T)$ between 17 K and 100 K, the activation energy can be estimated as the slope of the linear function

$$\ln(I(T) - I_0) = \ln(I_T) - E_{\alpha,a}\frac{1}{k_B T} \tag{5.6}$$

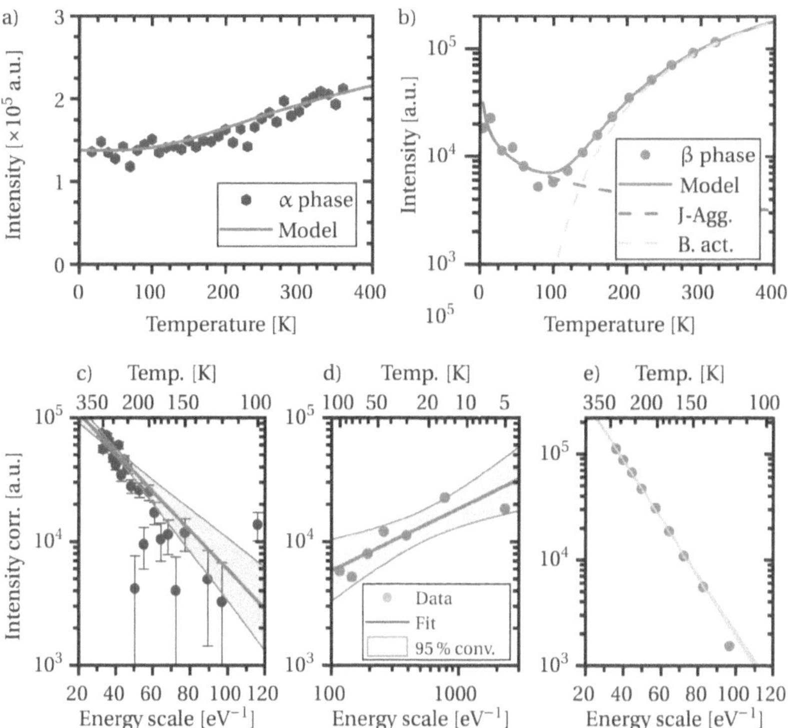

Figure 5.11.: Integrated PL intensities for the ZnPc a) α phase thin film and b) β phase crystal showing an overall increase in intensity with temperature. In c), d), and e) the (adjusted) intensities and corresponding linear fits, including 95 % confidence bands, are plotted over inverse thermal energy $1/k_bT$ abscissa for the α phase, the β phase low temperature, and the β phase high temperature emission, respectively.

over an $1/k_\mathrm{B}T$ abscissa. The temperature dependent part of the intensity data $I(T) - I_0$ plotted on a logarithmic scale over the energy scale abscissa $1/k_b T$ as well as the linear fit including a 95 % confidence band are depicted in figure 5.11 c). The fit yields an activation energy $E_{\alpha,\mathrm{a}} = (36 \pm 5)$ meV. By means of the parameters obtained by the fit and the mean value for the temperature independent emission rate, a curve corresponding to equation (5.5) is plotted in figure 5.11 a) tracing the observed emission behavior.

For the ZnPc β phase crystal, two independent processes govern the emission behavior. In the superradiance regime below 100 K the emission decreases with increasing temperature. According to equation (2.58) the emission rate and hence, the intensity from the radiative decay of a delocalized crystal exciton depends on the number of coherently excited monomers N_coh within the crystalline aggregate. Spano *et al.* [Spa10] derived that this number obeys the relation

$$N_\mathrm{coh}(T) = 1 + \left(\frac{4\pi\,|E_{12}|}{k_\mathrm{B}T} \right)^{\frac{d}{2}} \tag{5.7}$$

with E_{12} being the resonance interaction energy between two neighboring molecules (*c.f.* section 2.3.3) and d the dimensionality of the delocalization. In figure 5.11 d) the integrated intensity from 5 K to 100 K is plotted as a log-log plot against the $1/k_b T$ abscissa. Assuming $N_\mathrm{coh}(T) \gg 1$ for low temperatures, according to (5.7), $N_\mathrm{coh}(T)$ is described by a power law. As demonstrated in figure 5.11 d), fitting a linear function with fixed slope of $1/2$ to the data describes the low temperature emission well. The related dimensionality $d = 1$ agrees with the hypothesis of a J-aggregate preferentially stacking along the crystallographic [010] direction. Extrapolating the J-aggregate intensity power law to higher temperatures and subtracting the intensity contribution from the overall integrated intensity yields the neat intensity increase observed for temperatures above 100 K. This adjusted intensity is plotted in figure 5.11 e) on a logarithmic scale over the inverse thermal energy $1/k_b T$. Evidently, the data shows a linear behavior suggesting a Boltzmann activated process. A linear fit, also depicted in 5.11 with its 95 % confidence band, reveals an activation energy of $E_{\beta,\mathrm{a}} = (63 \pm 1)$ meV.

Hence, for both polymorphs a thermally activated PL intensity increase is observed. The respective activation energies suggest a phonon or molecular low energy vibrational mode mediating the depopulation of a dark state. A potential candidate for such a process could be the reverse intersystem crossing from a long living triplet state. The ZnPc triplet level lies approximately 1.12 eV above the S_0 state, as suggested by phosphorescence studies [VVR71]. This makes a repopulation of the molecular S_1 level at around 1.65 eV for the β phase and 1.74 eV for the α phase (*c.f.* thin film absorption in table 5.5) very unlikely, especially considering the observed activation energies. Moreover, Barbon *et al.* [BBF01a] concluded from electron paramagnetic resonance studies that the population of the triplet states by inter system crossing is suppressed in the solid state. As the energy condition for singlet fission, $2E_T \geq E_{S_1}$ [SM10][Cas18], is violated, triplet population by this mechanism is unlikely in ZnPc aggregates, also excluding an efficient triplet-triplet annihilation process with an underlying thermally activated triplet diffusion. This leaves molecular and inter molecular dark singlet states. For the α phase it can be rationalized that the reorganization of the inter molecular geometry upon excimer formation might be aided by the population of phonons performing a similar lattice deformation. In the following section 5.4.3 the vibrational spacing for the excited state inter molecular vibrations for the ZnPc α phase is estimated to about 25 meV to 30 meV, and hence lies in the same energetic range as the activation energy of $E_{\alpha,a} = (36 \pm 5)$ meV determined above. A similar comparison for the β crystal is missing and hence, proposing an explanation for the pronounced increase in intensity above 100 K with temperature is difficult without further research. As some reorganization after the excitation has been observed by comparing the spectral emission and absorption of single molecules in DMSO an intra molecular depopulation of a singlet dark state seems plausible.

5.4.3. ZnPc α Phase: Excimer Emission

As discussed in the previous section, the ZnPc α phase emission is of low intensity, largely red shifted and comprises a featureless line shape. The low intensity is consistent with the underlying H-aggregate packing as radiative transitions to the ground state are dipole forbidden. However, the large red shift and the missing structure in the spectral emission are peculiar as H-aggregates still show emission to excited vibrational modes of the ground state (*c.f.* 2.3.3). At low temperatures, assuming the Stokes shift similar to that of the 0-0 transition in the β phase of approximately 30 meV, the energetic distance of the emission maximum to the realted 0-0 crystal emission is about 380 meV. The vibronic transitions within this energy range known from ZnPc in solution (*c.f.* section 5.2.2) consist of excitations of mixed modes and higher ($i > 1$) vibrational levels. For an H-aggregate emission, radiative vibronic transitions to the first excited states of the respective vibrational modes of the ground state are possible and hence, should still be visible in the observed emission spectra, which is not the case. Thus, an H-aggregate emission does not sufficiently explain the observed emission behavior. However, the increase in the FWHM with temperature, the observed red shift, and the shape of the spectra are consistent with an excimer emission, which has been suggested before [Bal+09]. A semi-classical description of the excimer emission has been providen by Birks *et al.* [BKE68] for the example of pyrene crystal excimer emission. Their instructive approach, here referred to as the *Birks's approach,* will be adapted to describe ZnPc emission presented here.

The basic idea of the semi-classical description is explained with the help of figure 5.12. The distance between two neighboring molecules is given by the inter molecular coordinate q. By definition, in their electronic ground state $q = 0$ and for a decrease of inter molecular distance q increases to $q > 0$. Each molecule can be excited to its first excited state by absorbing a photon of energy M_0 which leads to the formation of a bound excimer state as the excitation is delocalized over two neighboring molecules and the inter molecular distance decreases by a value of q_e. The inter molecular potential can be approximated by a harmonic oscillator with the parabola's vertex located at position q_e and energy $D_e < M_0$, and with reduced mass μ and zero point energy E_0:

$$D(q) = \frac{2\mu E_0}{\hbar^2}\left(q - q_e\right)^2 + D_e. \tag{5.8}$$

The ground state potential which defines the final state into which the radiative transition takes places is given by

$$R(q) = R_0 q^2 \tag{5.9}$$

and is assumed to be continuous. This assumption has been justified by Williams and Hebbs [WH51] and later was formalized by Melvin Lax [Lax52]. It is based on the idea, that inter lattice site vibrations[4], *e.g.* in the here present case of a molecular crystal inter molecular vibrations, are usually of low frequency and if high vibrational states are populated, the Franck-Condon principle dictates a transition to the nuclear coordinate of the oscillator's turning points as at high quantum numbers the wave function of the harmonic oscillator is concentrated at these points. This means, that the final state coordinate and energy is reasonably well approximated by the ground state potential (5.9).

For simplicity, the excited state oscillator is assumed to be in the vibrational ground state with energy $D_e + E_0$ and the derived principle will later be extended to the case of higher vibrational levels. If the final state is located at the inter molecular coordinate $q = q^*$ the energy of the emitted photon is given by

$$E(q^*) = (D_e + E_0) - R(q^*). \tag{5.10}$$

The probability of the two molecules constituting the excimeric state to be at the inter molecular coordinate q^* is given by the squared absolute of the excited state oscillator

---

[4]As the theory by Williams and Hebbs has been generally derived and applied to the emission band of the ionic crystal KCl activated with Thallium, this somewhat bulky term hints to the applicability for many systems.

wave function $\left|\Psi(q^*)\right|^2 \mathrm{d}q$ in the infinitesimal interval $\mathrm{d}q$. The transition spectrum of the emission $P(E(q^*))\mathrm{d}E$ is then related to the spatial probability distribution as

$$P\left(E(q^*)\right)\mathrm{d}E = \left|\Psi(q^*)\right|^2 \mathrm{d}q. \tag{5.11}$$

The emission energy is related to the inter molecular coordinate by equation (5.10). Using the inverse function with the appropriate Jacobian $\mathrm{d}q/\mathrm{d}E$, the transition spectrum of the excimer emission can be calculated from the excited state oscillator wave function as

$$P(E)\mathrm{d}E = \left|\Psi\left(q(E)\right)\right|^2 \frac{\mathrm{d}q}{\mathrm{d}E}\mathrm{d}E. \tag{5.12}$$

The resulting emission transition spectrum for radiative transitions from the vibrational ground state of the excimer state is shown on the vertical energy axis on the right side of figure 5.12.

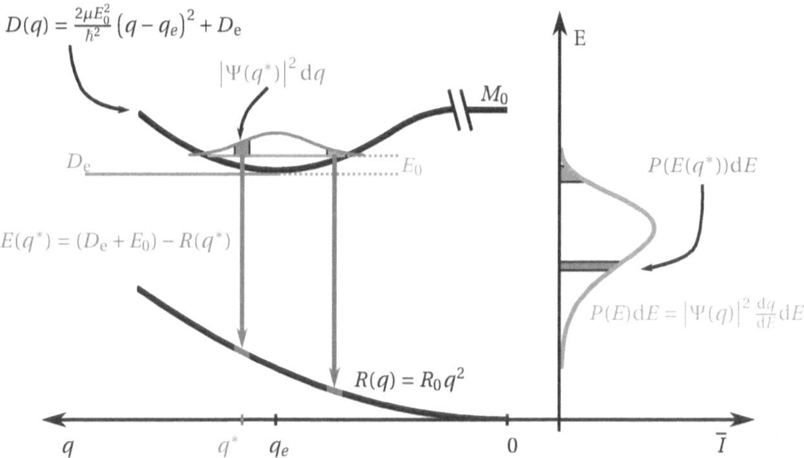

Figure 5.12.: Semi-classical approach to calculate the excimer emission spectra. The q abscissa of the inter molecular distance points in positive direction from right to left by definition. The excited state is assumed to behave like a harmonic oscillator at energy D_e emitting into a quasi continuous repulsive ground state potential $R(q)$. The spectral density as a function of the emission energy $P(E)\mathrm{d}E$ is then directly related to the spatial probability distribution of the harmonic oscillator given by the squared absolute of the vibrational dimer wave function $\left|\Psi(q)\right|^2\mathrm{d}q$. The figure is based on [WH51] and [BKE68].

Extending the above concept to the n-th oscillator mode leads to an emission transition spectrum $P_n(E)\mathrm{d}E$ determined by the wave function of the respective state $\Psi_n(q)$. The resulting emission energy is derived from equation (5.10) by substituting the oscillator ground state energy with the energy of the n-th vibrational mode $(2n+1)E_0$. If the excimer is treated at a finite temperature T the excited state oscillator is occupied according to a Boltzmann distribution $B(n,T)$ for the states n. The resulting transition spectrum is described by the Boltzmann weighted sum of the individual transition lines

$$P(E;T)\mathrm{d}E = \sum_n B(n,T)P_n(E)\mathrm{d}E = \sum_n B(n,T)\left|\Psi_n\left(q_n(E)\right)\right|^2\frac{\mathrm{d}q_n}{\mathrm{d}E}\mathrm{d}E \qquad (5.13)$$

where $q_n(E)$ is the inverse function of (5.10) for the excited state n, respectively. For an excited state described by a harmonic oscillator the analytical form of $P_n(E)\mathrm{d}E$ is sketched in appendix B.2. The analytical expressions of $P_n(E)\mathrm{d}E$ for n ranging from 0 to 5 are presented by equations (B.15) to (B.20).

To characterize the excited state potential Birks *et al.* [BKE68] used the properties of the Slater sum of a harmonic oscillator [AM91] to derive a temperature dependent expression for the FWHM of the emission spectra:

$$\Delta P(T) = \Delta P_0 \coth\left(\frac{\theta}{T}\right)^{\frac{1}{2}}. \qquad (5.14)$$

Here, θ is the temperature equivalent of the zero point energy of the excited state oscillator $k_\mathrm{B}\theta = E_0$. A semi-classical emission transition spectrum by Keil [Kei65a] implying the same temperature dependency for the FWHM has recently been used to explain the emission spectra of CT states based on inter molecular vibrations [TBV20]. An important restriction of equation 5.14, besides its semi-calssical approximation and its excited state potential being independent of temperature, is its validity for symmetric emission profiles only. This is the case if the ground and excited state have the same potential as assumed by Keil [Kei65a] or if the Jacobian does not deform the wave functions Ψ_n of the harmonic oscillator which are added up in (5.13) which is valid for, *e.g* the linear approximation of the ground state potential of Williams [WH51].

The key assumption for the description of the ZnPc α phase emission is that the excimer luminescence is governed by a single inter molecular vibration which can be approximated by a harmonic oscillator with a temperature independent potential. Based on the molecular packing in α crystal structure, in the (010) plane a preferred displacement along the inter molecular short crystallographic [100] axis can be assumed (*c.f.* figure 5.6 c)). However, knowledge on the exact relation to the harmonic potential is not necessary for a mathematical description. As becomes evident by the presented data in figure 5.10, the line shapes of the transition spectra are asymmetric for low temperatures, which is why equation (5.14) can only provide an estimation of the excited state potential. The FWHM variation of the emission transition spectra with temperature is plotted against $\sqrt{(T)}$ for better presentation of the low temperature region in figure 5.14 a). A fit by equation (5.14) yields $\Delta P_0 = (148 \pm 1)\,\text{meV}$ and $\theta = (171 \pm 5)\,\text{K}$ corresponding to a zero point energy of $E_0 = (14.7 \pm 0.4)\,\text{meV}$ defining the harmonic potential of the excimer state by its vibrational ground state energy.

In the next step, equation (5.13) with $n \in \{0,\dots,5\}$ is used to fit the obtained emission data. The zero point energy of 14.7 meV is applied for the excited state oscillator over the whole temperature range while the energetic offset D_e, the inter molecular displacement q_e, and the ground state potential parameter R_0 are chosen as free, temperature dependent parameters. This proved to be necessary as a temperature independent ground state potential and displacement were able to reproduced the shape of the spectra but not it's energetic position. The emission data (solid line, purple), the fitted emission (solid line, orange), and the individual emission profiles P_n (dotted line, pink) are shown in figure 5.13 for selected temperatures.

The fitted spectra match the observed emission transition spectra very well over the whole temperature range. For low temperatures, the overall line shape as well as the position of the emission maximum deviate from the experimental data. This shortcoming is attributed to the mismatch between the assumed excited state potential given in terms of its zero point energy and the actual one. As the condition of a symmetric emission profile is violated for equation (5.14) at low temperatures. This leads to a mismatch between the FWHM data and the fit curve. With rising temperature, the effect of the mismatch steadily vanishes. At high temperatures ($\geq 300\,\text{K}$) the energy of emitted photons can surpass the energy level of the excimer's vibrational ground state $D_{E_0} = D_e + E_0$. As for n-th Jacobian $\mathrm{d}q_n/\mathrm{d}E$ a singularity is found at D_{E_n} (*c.f* equation (B.11)), the semi-classical emission profile of each vibrational level $P_n(E)\mathrm{d}E$ is only defined for $E < D_{E_n}$. Hence, the full profile described in equation (5.13) is restricted to $E < D_{E_0}$, as is indicated by the cutoff of the fitted emission profile at about $D_0 + E_0 \approx 1.55\,\text{eV} + 0.015\,\text{eV} = 1.565\,\text{eV}$ in figure 5.13 (only deviating for $T > 330\,\text{K}$). At 17 K, the emission is dominated by the excimer's vibrational ground state represented by the P_0 line shape. With rising temperature, the occupation of the subsequent vibrational levels increases and their radiative relaxation to the electronic ground state contributes to the overall emission. At about 120 K, the emission of the first excited state P_1 becomes relevant to the overall emission. The line shape of the latter is caused by a distorted squared vibrational wave function with its mirror symmetry around the node distorted by the curved ground state potential. Above 280 K, the next excimer vibrational level starts contributing and the high energy side is increasingly dominated by the P_2 line shape. Higher vibronic transitions seem not to be contributing signifi-

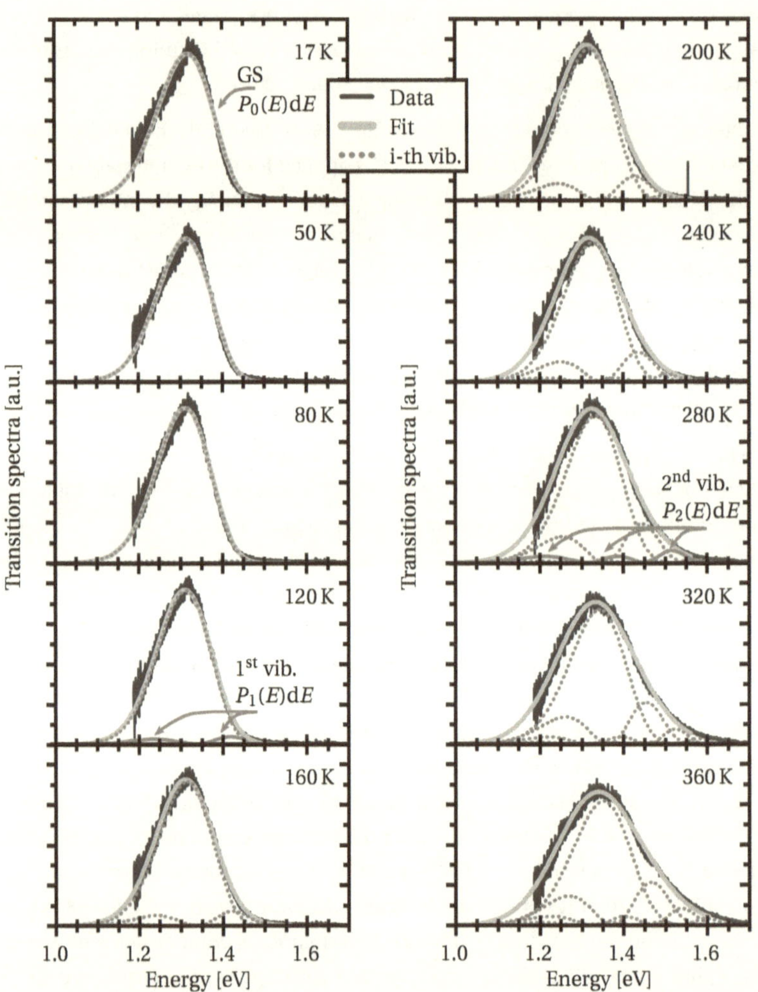

Figure 5.13.: Semi-classical fit of excimer emission spectra for different temperatures. With increasing temperature the emission from higher vibrational states significantly contributes to the spectra. The individual contributions are indicated by the dotted pink lines and each contribution is labeled in the spectra according to its first appearance.

cantly to the observed spectra. As no part of the measured emission spectra is missing and the contribution of each vibrational level is determined by the Boltzmann occupation of the excited state oscillator, the estimated zero point energy of 15 meV proves to be a good overall approximation.

The obtained fit parameters are illustrated in figure 5.14 b) and c). The electronic energy gap D_e is plotted against the temperature in 5.14 b). For $T \leq 330\,\mathrm{K}$ its value is constant at $(1.547 \pm 0.011)\,\mathrm{eV}$ (smallest obtained error of 2 meV). This result agrees with the expectation that the energy of an electronic state is mainly independent of the system's temperature. At high temperatures, fit function (5.13) loses its validity for the present high energy emission above 1.56 eV and the minimization algorithm is forced to blue shift the whole emission spectra to compensate for the encountered singularity in the Jacobians by increasing the energetic position D_0 of the excited state.

Figure 5.14 c) depicts the ground state potential parameter R_0 and the inter molecular displacement q_e as function of temperature. As the general shape of the spectra is mostly determined by the excited state wave functions and the excited state potential, R_0 and q_e determine the energy scale and the asymmetry of the emission line shape and hence, are correlated as can be seen by their opposite progression. For low temperatures, where a strong asymmetric emission is observed both quantities deviate from their otherwise only slightly varying values between 50 K and 150 K. With rising temperature, an increasing variation in both parameters is evident up to 330 K, where the break down of the algorithm validity is indicated by a discontinuous change in both parameters. The monotonous change within the valid temperature range above 150 K can be attributed to the expected lattice expansion with increasing temperature. One could argue that such a change in the lattice constant will influence the excited state potential as well. However, the excimer potential is mainly constituted by the binding character of the excited state wave function delocalized over the two molecules. Therefore, while the surrounding crystal lattice changes, the excimer potential, as a first approximation, can be regarded as constant with temperature.

In figure 5.14 d) the zero point energy corresponding to the ground state oscillator is plotted for the purpose of comparability with the excimer oscillator. Evidently, the zero point energy is smaller than the estimated 15 meV of the excited state potential for all temperatures, implying a tighter inter molecular binding for the excited state due to the additional resonance energy exchange by exciton delocalization. The Huang-Rhys parameter S, referred to the ground state quantization, is displayed as well (red, right y-axis). According to its definition given by equation (2.38), S equals the number of vibrational quanta needed for the displacement q_e, showing that the "average transition" has its final state at highly excited vibrational levels of the electronic ground state. This

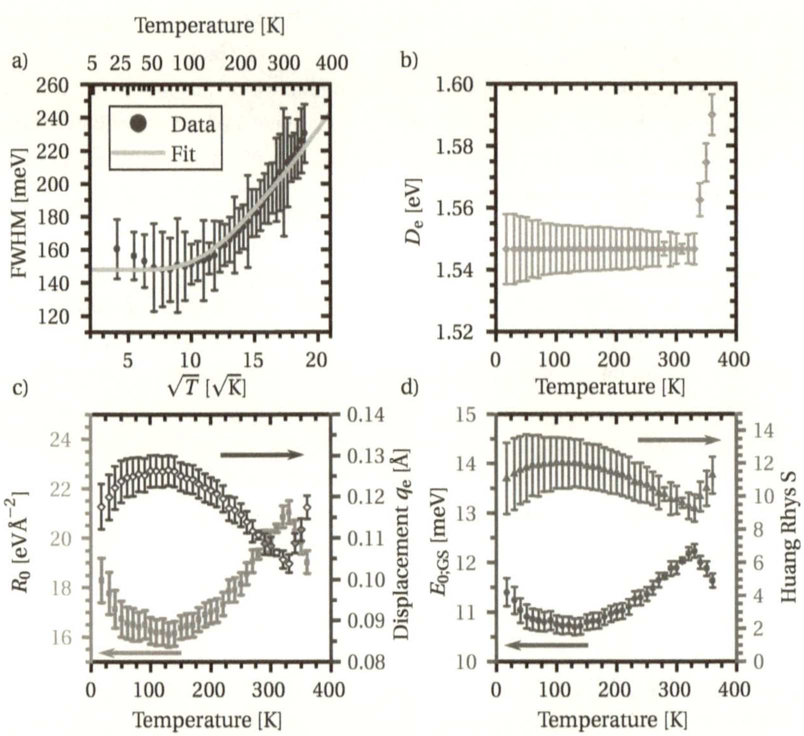

Figure 5.14.: a) FWHM of the ZnPc α phase emission together with fitted curve according to equation (5.14). Obtained fit parameters of the excimer fit according to equation (5.13): b) Electronic energy gap D_e, c) ground state potential parameter R_0 (pink, left y-axis) and inter molecular displacement q_e (brown, right y-axis). d) Zero point energy of the ground state oscillator (blue, left y-axis) and Huang-Rhys parameter S (red, right y-axis) with referred to the ground state.

implies that the assumption of a distinct inter molecular coordinate of the final state is justified as the precondition of a wave function localized at the turning points of the oscillator is fulfilled for such high energy vibrational levels.

The semi-classical model of an excimer state describes the observed ZnPc emission reasonably well over a broad temperature range between 20 K and 360 K and thus, corroborates the fundamental assumption of an excimeric emission. The passage of the molecular excited state into an excited dimer state including inter molecular relaxation, as well as the emission attendent phononic lattice disturbances lift the restriction of dipole forbidden transitions to the Γ point of the exciton dispersion imposed by the H-aggregate packing, as the momentum mismatch is compensated by the involved molecular motions.

The Birk's approach for describing molecular excimers is, of course, only an approximation, as in general the ground state needs to be quantized as well and the emission spectrum is given by discrete vibronic transitions with their intensities determined by the respective Franck-Condon factors. In fact, the lack of computational power at the time of the development was the main reason for the development of a semi-classical description, as Williams and Hebb state [WH51]: "A serious problem arises [...] because transitions occur to high vibrational levels in the final state. [...] Hermite polynomials of this order are not available in tabulated form and their computation would involve enormous labor." This is no longer a problem as numerical evaluation of Franck-Condon integrals can be carried out quickly by modern computers.

The discrete quantum mechanical transition spectrum is given by equation (2.49). As the excimer and ground state potential are not of the same shape, the Franck-Condon factors can not be formulated as an analytical expression comparable to equation (2.39), but need to be evaluated numerically. The Franck-Condon factor describing a vibronic transition from the k-th vibrational excimer state to the j-th vibrational ground state is given by

$$F_{kj}\left(q_e, E_{0;EX}, E_{0;GS}\right) := \left|\langle j \mid k \rangle\right|^2 = \left|\int_{-\infty}^{\infty} \psi_k\left(q - q_e; E_{0;EX}\right) \psi_j\left(q; E_{0;GS}\right) dq\right|^2 \quad (5.15)$$

with the inter molecular displacement q_e and the zero point energies $E_{0;EX}$ and $E_{0;GS}$ of the excimer and ground state, respectively, defining the potentials' strength. With equation (5.15), the electronic energy gap D_e and a the Gaussian line shape function

(5.2), the full emission spectrum normalized to the electronic transition rate (2.49) becomes

$$\overline{I}\left(E;\, T,\, q_{\mathrm{e}},\, E_{0;\mathrm{EX}},\, E_{0;\mathrm{GS}},\, D_{\mathrm{e}},\, \sigma\right) =$$
$$\sum_{k}\sum_{j} B(k, T)\cdot F_{kj}\left(q_{\mathrm{e}},\, E_{0;\mathrm{EX}},\, E_{0;\mathrm{GS}}\right)\cdot\Gamma\left(E;\, D_{\mathrm{e}} + (2k+1)E_{0;\mathrm{EX}} - (2j+1)E_{0;\mathrm{GS}},\, \sigma\right) \quad (5.16)$$

with σ being a static energetic disorder parameter determining the line width of the individual transitions. A temperature independent electronic energy gap D_{e} and energetic disorder σ will be assumed. Additionally, it is reasonable to approximate the excited state potential to be constant with temperature as well. A global fit procedure of the python *LMFIT* package [New+14] which minimizes a single residual function consisting of all experimental and simulated spectra by a least square method has been applied. By means of this approach the global, *i.e.* temperature independent, parameters and the temperature dependent parameters are determined simultaneously. This enables a detailed analysis of the emission spectra. Details on the fitting procedure can be found in appendix B.3. The resulting transition spectra are shown for representative temperatures in figure 5.15. The emission of the first five excimer vibrational states are color coded and the stick emission spectra indicates the energetic position and intensity of each vibronic transitions to the electronic ground state.

First of all, it can be noted that the Franck-Condon model does describe the observed emission spectra very well, although, some deviations from the data occur. For low temperatures ($T < 80\,\mathrm{K}$), the emission maximum is slightly red shifted with respect to the experimental data and the overall line shape shows a slightly different course. For the high temperatures ($T \geq 320\,\mathrm{K}$), the high energy side is underestimated. Interestingly, both deviations are found in the Birk's approach as well hinting at a shortcoming in the general assumptions but not in the simplifications in the semic-lassical description. As previously discussed, the low temperature emission is dominated by the excimer's vibrational ground state which is assumed to be constant over temperature and hence, was treated as a global fit parameter. This could lead to some deviations for at the extremes of the observed temperature range while the overall data is reproduced sufficiently well. An additional point to discuss are the key assumptions of a single inter molecular vibration governing the transitions and of a harmonic ground and excited state potential. The stacking of the ZnPc molecules along the [100] unit cell direction is not exactly cofacial, but rather slip-stacked, *i.e.* the molecules' centers are slightly shifted with respect to each other, which could lead to a shearing motion which superimposes the assumed displacement and vibration along the inter molecular axis. Thus, the assumption of a single mode may very well be oversimplified leading

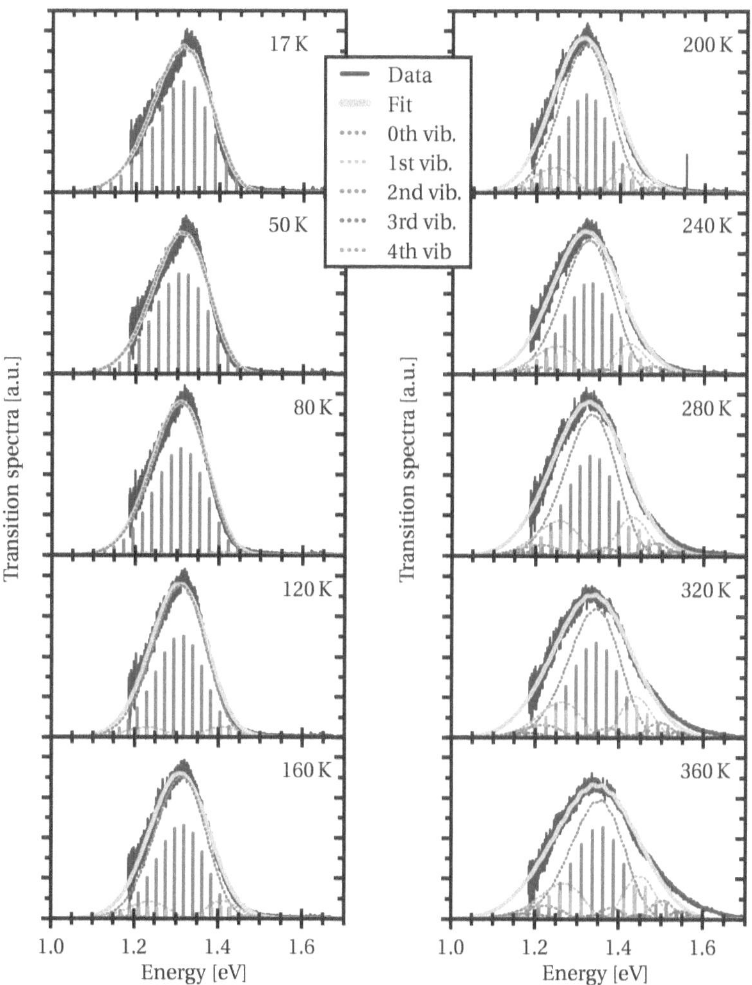

Figure 5.15.: Quantum mechanical fit of the ZnPc excimer emission spectra at different temperatures. With increasing temperature, the emission from higher vibrational states significantly contributes to the spectra. These individual contributions are indicated by the dotted curves, and each vibronic transition and its amplitude is depicted by a vertical line and its height.

to deviations for very low and very high temperatures. As the approximation of a harmonic potential is usually only valid for low quantum numbers, the anharmonicity will lead to the formation of a dominant high energy tail, as discussed in section 2.3.2.

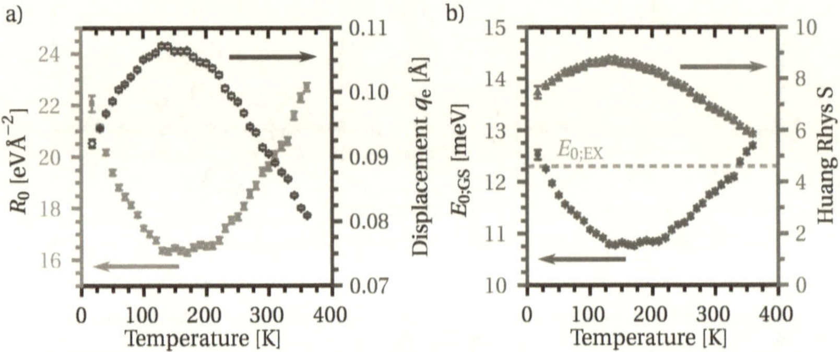

Figure 5.16.: a) Ground state potential parameter R_0 (pink, left y-axis) and inter molecular displacement q_e (brown, right y-axis). b) Zero point energy of the ground state oscillator (blue, left y-axis) and Huang-Rhys parameter (red, right y-axis) with regard to the ground state. The zero point energy of the excited excimer state is indicated by the dashed green line.

The fit yielded an energetic disorder of $\sigma = 13.3\,\text{meV}$, an electronic energy gap of $D_e = 1.489\,\text{eV}$ and an excited state zero point energy of $E_{0;EX} = 12.3\,\text{meV}$. For all parameters the obtained standard error was below 1 % and hence, much lower than the accuracy expected in the model's limitations. The excited state zero point energy as well as the electronic band gap are smaller compared to the parameters of the Birk's approachs. In figure 5.16 a) the progression of the ground state potential and the inter molecular displacement are shown as functions of temperature. The ground state potential is found to be of similar range compared to the semi-classical fit, however, showing a larger spreading at temperatures below 100 K. The inter molecular displacement is found at overall lower values. In contrast to the Birk's approach, for which at high temperatures the singularities in the Jacobian lead to a discontinuity in the progression of the fit parameters q_e and R_0 with temperature, the two parameters do not experience any discontinuity and show the same trend as discussed previously for the semi-classical model over the whole temperature range. In figure 5.16 b) the Huang-Rhys parameter with respect to the ground state oscillator and the zero point energy of the ground state oscillator are shown. The fixed zero point energy of the excited state is indicated by the dashed green line. While for the semi-classical description the ground state potential was always characterized by a lower zero point energy than the excited state, the zero point energies cross for low and high temperatures. Although overall slightly lower than for the semi-classical model, the Huang-Rhys factors are still high for all temperatures, again justifying the main assumption of the semi-classical approach.

The quantum mechanical description and the resulting fit lead to two main conclusions. At first, it confirms the validity of the semi-classical approach. For inter molecular vibrational energies in the range of 10 meV to 50 meV, *i.e.* typical phonon energies [Sha+14], and a sufficient inter molecular displacement the assumption of a continuous ground state potential is a valid simplification yielding to comparable results to a quantum mechanical treatment. This is especially helpful if an analytical expression of the transition spectrum is needed. However, caution is advised if the emission energies are close to the electronic energy gap for high temperatures, as the singularity in the Jacobian restricts the validity of the transition spectrum. Secondly, the relation between the ground state and excited state strength determine the appearance of the emission line shape. In figure 5.16 the emission transition spectra for different ground state potentials and a fixed excited state potential are shown for different temperatures. The relation of the two potentials is given in terms of their respective zero point energies. For a steeper ground state potential (left, $E_{0;GS} > E_{0;EX}$) the spectra show a strong asymmetry towards the low energy side of the emission. With rising temperature, the intensity of the emission maximum decreases quickly and the asymmetry in the line shape becomes stronger. For the common assumption of equal potentials in the ground and the excited state (center, $E_{0;GS} = E_{0;EX}$), a symmetric Gaussian emission profile is found. With rising temperature the FWHM maximum becomes larger and, at the same time, the intensity of the emission maximum quickly decreases. Under these circumstances, an analytical solution of the Franck-Condon factors is given by equation (2.39). In the case of a steeper excited state potential, the emission is comparably narrow, albeit, the FWHM is still around 100 meV. With rising temperature a slight asymmetry is found at the high energy side. The decrease in the emission maximum is not as pronounced as for the other cases and a slight red shift with rising temperature is found.

The semi-classical as well as the quantum mechanical description of the excimeric state is well suited to describe the present emission of the ZnPc α phase suggesting excimer formation to be the dominant emission mechanism. This mechanism is closely linked to the molecular packing within the unit cell. The H-aggregate packing prohibits a radiative exciton decay. As the inter molecular distance is short, the relaxation of two neighboring molecules into an inter molecular potential minimum becomes a favorable relaxation pathway for the excited state. The inter molecular displacement

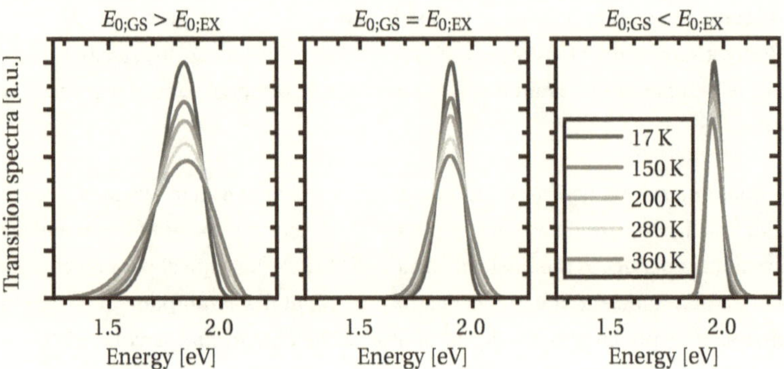

Figure 5.17.: Influence of the relation between ground state ($E_{0;GS}$) and excited state potential ($E_{0;EX}$) and the resulting emission transition spectra for different temperatures.

enables radiative transitions in the H-aggregate as the transition is accompanied by phonon generation providing the necessary conservation of momentum.

5.4.4. ZnPc β Phase: High Temperature Frenkel Emission

As discussed previously in section 5.4.2 the emission of the ZnPc β phase can be coarsely divided into two temperature regimes. While for temperatures below 100 K a superradiant behavior emerging from exciton delocalization in a J-aggregate confirmation leads to a narrow dominant emission line, at higher temperatures of above 180 K, a typical Frenkel type exciton emission is observed. The two regions can be clearly distinguished by their respective integrated intensity evolution with temperature shown in figure 5.11 b). While for low temperatures the J-aggregate emission dominates, the intensity at higher temperatures is governed by a temperature activated process. At an intermediate temperature region, between 100 K and 180 K, both processes coexist and contribute to the overall intensity. As mentioned, the superradiant low temperature emission will be subject to a separate examination below(*c.f.* section 5.6.3). Hence, in this section, the high temperature emission is analyzed by examining the photoluminescence above the intermediate temperature range.

Similar to the thin film PL the crystal shows a Frenkel type exciton polaron emission with the main 0-0 transition peak accompanied by progressed vibronic transitions to molecular vibrational levels (*c.f.* section 2.3.3). A Franck-Condon fit based on the solvent emission spectra in DMSO, as discussed in section 5.2.2, is employed to test the relation to the single molecule excited state emission. Using equation (5.3) with three vibrational modes of fixed energies $E_{vib,\alpha} = 78\,\text{meV}$, $E_{vib,\beta} = 142\,\text{meV}$, $E_{vib,\gamma} = 180\,\text{meV}$, the 0-0 transition energy E_0, the energetic disorder parameter σ, and the respective Huang-Rhys parameters S_α, S_β, and S_γ as free parameters the data is fitted between 180 K and 320 K. The results are shown in figure 5.18. Table 5.6 lists the deduced fit parameters together with their errors as obtained by the fitting procedure omitting those smaller than the reasonable precision of the simulation

For high temperatures, *i.e.* between 260 K and 320 K, the fit is in good agreement with the data. At 230 K slight deviations around 1.5 eV are visible caused by an additional peak evolving at the low energy shoulder of the initial vibronic hump. Below 200 K, the deviation of the resulting fit (dotted grey line) from the experimental data in the region of the vibronic progressions becomes larger and the model of the proposed vibrational modes seems no longer valid.

T [K]	E_0 [eV]	σ [meV]	S_α	S_β	S_γ
320	1.593	39.6	0.325 ± 0.006	0.106 ± 0.006	0.134 ± 0.004
290	1.594	36.6	0.340 ± 0.004	0.144 ± 0.004	0.130 ± 0.003
260	1.595	34.0	0.364 ± 0.003	0.194 ± 0.004	0.118 ± 0.003
230	1.596	31.9	0.395 ± 0.003	0.244 ± 0.004	0.105 ± 0.003
200	1.598	28.7	0.432 ± 0.004	0.274 ± 0.005	0.120 ± 0.004
180	1.560	28.0	0.503 ± 0.005	0.315 ± 0.006	0.155 ± 0.005

Table 5.6.: Fit parameters of the employed Franck-Condon fit: 0-0 transition energy E_0, energetic disorder parameter σ, and Huang-Rhys parameters S_α, S_β, and S_γ of the three vibronic transitions (fading values are of no significance for the discussed model). Errors are obtained from fitting procedure. Omitted error values are smaller than the precision of the fit.

For $T \leq 200$ K the vibronic progression can no longer be reproduced and hence, the obtained Huang-Rhys parameters are without relevance as indicated by the fading text color. The 0-0 transition is reproduced reasonably well which is why E_0 as well as the energetic disorder σ are considered to be correct as they can be determined exclusively from the 0-0 main transition.

Comparing the data in figure 5.18 with the single molecule emission in DMSO in figure 5.2, it is noticeable that the prominent second peak about 170 meV below the 0-0 transition is missing in the crystal emission. Instead, only a shoulder at the respective energy is present. For temperatures above 230 K this results from a significant increase in the effective vibrational coupling of the α and β mode to the electronic transitions. Comparing the Huang-Rhys parameters to the ones of the single molecule emission listed in table 5.2 highlights a (54 ± 3) % to (89 ± 2) % increase for S_α and (30 ± 8) % to

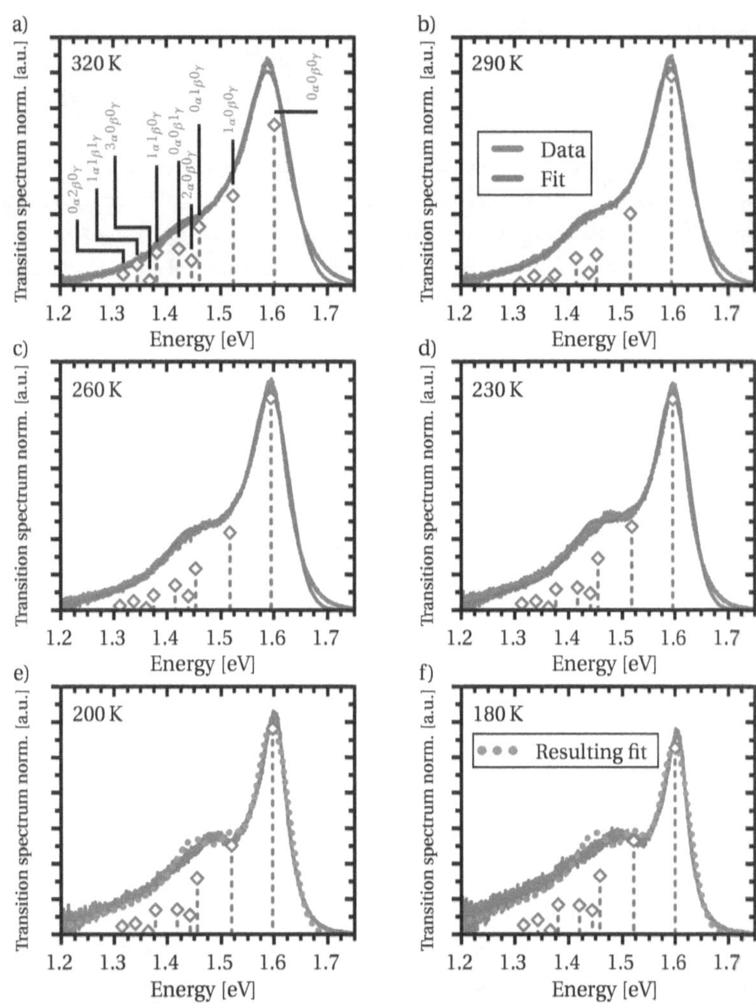

Figure 5.18.: High temperature emission of znPc β phase crystal evaluated with a Franck-Condon model for different temperatures. Stick spectra (only shown up to 1.3 eV) show contributions of different vibronic transitions. From 320 K to 260 K, as shown in a), b), and c), a model containing three vibrational modes describes the observed emission well, while in d) at 230 K small deviation between measured and simulated data are visible. Below 200 K, in e) and f), the assumed vibrational modes do no longer resemble the observed PL and the resulting fit (dotted grey line) significantly deviates from the experimental data.

$(201 \pm 9)\,\%$ increase for S_β while the variation in S_γ is comparably small at values between $(-8 \pm 3)\,\%$ and $(17 \pm 4)\,\%$. It has to be mentioned that the omission of a self absorption correction leads to an overestimation of the extracted Huang-Rhys parameters as discussed in section 5.4.1. However, the striking difference between the single molecule emission and the crystal PL cannot solely be explained by self absorption and the increase in vibronic coupling is a real effect of the crystal environment rather than an artifact by self absorption. Such an increase in vibronic coupling induced by the crystalline environment has also been reported for other molecular systems [Mül+20].

Interestingly, at lower temperatures, the emission shows a behavior known for H-aggregates, *i.e.* a dominant vibronic progression with a low 0-0 transition intensity the latter decreasing with decreasing temperature [Spa10]. This emission contradicts the observed J-aggregate molecular packing found in the ZnPc β crystal structure and the corresponding emission for high and very low ($T \leq 100\,\mathrm{K}$) temperatures. To the author's best knowledge no phase transition has been reported for ZnPc at low temperatures, however, phthalocyanines in general are know to express highly spatially anisotropic thermal expansion and contraction [UWD43]. Thus, the presented spectroscopic data suggests that the crystal packing continuously changes from a J-aggregate to an H-aggregate packing before changing back to a J-aggregate packing leading to the observed emission behavior below 100 K.

In conclusion, the preceding analysis indicates that the high temperature single crystal emission is determined by two crystallographic aspects. First, the J-aggregate packing leads to a dipole allowed 0-0 transition, giving rise to the observed strong Frenkel exciton polaron emission. In addition, the second molecule in the unit cell (*c.f.* figure 5.6 c) and d)) sterically hinders an excimer formation along the short [010] crystallographic axis, leaving the Frenkel exciton decay as the sole radiative emission pathway. This leads to a much higher emission quantum yield of at least a factor of five (*c.f.* section 5.5.1) and higher photon energies in the ZnPc β phase emission compared to the α phase, unambiguously demonstrating a crucial influence of the molecular packing on the photophysical processes in molecular aggregates.

5.5. Characterization and Application of Phase Transition Kinetics

So far, the interplay between the crystal structure of two common ZnPc polymorphs and their aggregate photophysics has been discussed. Analyzing their distinct emission characteristics in relation to their crystal structure offers insights into the underlying photophysical and excitonic processes. The spectral differences in the emission profiles of the two crystallographic phases allows for identification of the respective crystal structure by its characteristic emission. This correspondence enables tracking of the structural transition dynamics between α and β phase over time. The large spectral separation in combination with a thermally induced phase transition enables the spectral tuning of a single organic material without the need for elaborate synthetic efforts on the molecular structure. This opens the perspective to design dual emitting OLEDs based on a single active molecular compound.

After characterizing the phase transition kinetics by means of the distinct emission signatures , the gained insights are brought into application in three prototypical diodes based on the α, β, and a mixed phase demonstrating the selective tuning of the respective electroluminescence (EL).

The results presented in the following sections have been published in [Ham+19]. Data analysis as well as the interpretation is taken from the above publication if not stated otherwise.

5.5.1. Deducing Phase Transition Kinetics by Photoluminescence

The phase transition was optically monitored using the thermal annealing setup described in section 3.4.2 and measuring the PL at a 685 nm laser excitation. Spectra were recorded continuously with the $150\,\mathrm{lines\,mm^{-1}}$ diffraction grid in one minute time steps with a 30 s acquisition time. Thin film samples were prepared, as described in section 5.3.2, by vacuum sublimation of purified (twofold gradient sublimation) ZnPc at a deposition rate of $0.17\,\mathrm{\AA\,s^{-1}}$ up to a nominal film thickness of 30 nm. Silicon wafers covered with a thermally grown SiO_2 were used as substrates. The samples were slowly heated up to 250 °C. During the heating procedure PL spectra were recorded continuously to monitor the emission profile and exclude phase transition before the final temperature was reached. At 250 °C the temperature was kept constant and the emission behavior was monitored over 120 min until no further spectral changes were observed. Exemplary spectra at 15 min, 35 min, 50 min, and 100 min are shown in figure 5.19 and the evolution of the emission is depicted in figure 5.20 a).

At the beginning of the heating cycle the spectra clearly show the broad α phase excimer emission with its maximum at around 1.4 eV (figure 5.19: 15 min). Compared to the emission at 90 °C (corresponding to 360 K), depicted in figures 5.13 and 5.14, the FWHM maximum is increased further as expected for an excimer emission at higher temperatures. After about 30 min, the high energy side is superimposed by an emerging peak at 1.6 eV, matching the β phase thin film emission. This new component keeps growing in intensity over the next 60 min until the thin film PL has been completely converted to that of the neat β phase and no further changes occur, indicating the completion of the phase transition. The overall intensity is increased by a factor of five, caused by the higher quantum efficiency of the Frenkel exciton emission compared the α phase excimer emission, the latter to be prone to thermally activated non radiative decay channels [WPW85].

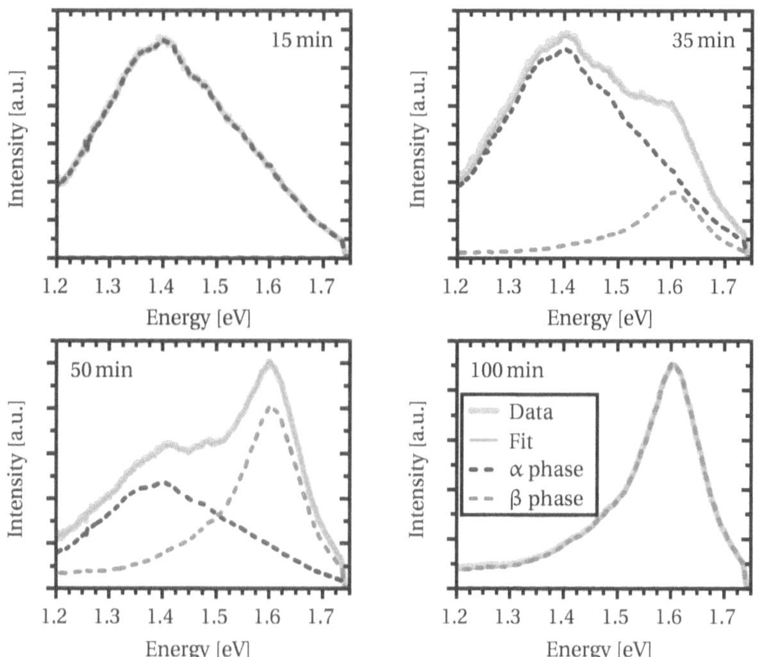

Figure 5.19.: Emission spectra at the beginning (15 min) and the end (100 min) of the ZnPc α → β phase transition with two intermediate spectra (35 min and 50 min) constituting of the emission of both polymorphs. Fit curve (orange line) and spectra of the neat polymorphs scaled by their respective volume part (α: purple, β: red) shown in all spectra. Data of 15 min and 100 min has been previously published as supplementary information in [Ham+19].

Under the assumption that the emitted intensity of either polymorph is directly proportional to its volume fraction in the probed region, a quantitative analysis of the phase transition kinetics is possible. Averaging 15 spectra at the beginning and the end of the 120 min annealing procedure yields reference spectra of the α phase β phase,

$S_\alpha(E)$ and $S_\beta(E)$, respectively. With the present volume fraction p of the β phase being present at any given time of the phase transition the composed spectra is given by

$$S(E; p) = p \cdot S_\beta(E) + (1 - p) \cdot S_\alpha(E) \tag{5.17}$$

with $(1 - p)$ as the corresponding volume fraction of the α phase. By fitting equation (5.17) to each spectra recorded during the phase transformation the respective volume part of each polymorph can be traced over the whole conversion process. In figure 5.19, the extracted contributions of each phase to the overall emission are shown as dashed lines (α: purple, β: red) and the resulting fit curve, depicted in orange, shows an excellent agreement with the data. The volume parts of each phase, p and $(1 - p)$ are plotted as function of time in figure 5.20 b). The discontinuity after about 75 min can be attributed to a slight change of the sample position, leading to a sudden increase in intensity which can also be seen in the related spectra in figure 5.20 a).

Supposing an isothermal phase transition, the data can be analyzed by the Johnson-Mehl-Avrami-Kolmogorov (JMAK) model (*c.f.* section 2.2.3). The volume fraction of the resulting phase is given by equation (2.20) simplifies to

$$p(t) = 1 - e^{-V_e(t)} = 1 - e^{-Bt^n} \tag{5.18}$$

if a constant nucleation rate is assumed. The constant B unites the growth rate, the shape factor, and the nucleation rate in an effective rate constant determining the timescale of the transformation. The *Avrami exponent* $n = d + 1$ depends on the dimensionality d of the process, *i.e.* $d = 3$ for a polyhedral, $d = 2$ for a plate like, and $d = 1$ for a needle like growth as given by equations (2.23) to (2.25). Equation (5.18) can be converted to

$$\log\left(-\ln\left(1 - p(t)\right)\right) = \log(B) + n \cdot \log(t) \tag{5.19}$$

yielding a linear function with the Avrami exponent n as the slope. In figure 5.21 the volume part $p(t)$ has been transformed according to the left side of equation (5.19) and is plotted over a logarithmic time scale. A 1D, 2D, and 3D JMAK model has been used to fit the data between 20 min and 100 min, and the resulting curves are shown as dashed lines in figure 5.19 as well. Clearly, the data suggests a polyhedral growth mechanism driving the phase transition, despite the needle like morphology of the resulting β phase crystallites (*c.f.* figure 5.7 b)). The resulting progression of p over time is shown in figure 5.20 b) as a dashed blue line, reproducing the observed transition kinetics in good agreement.

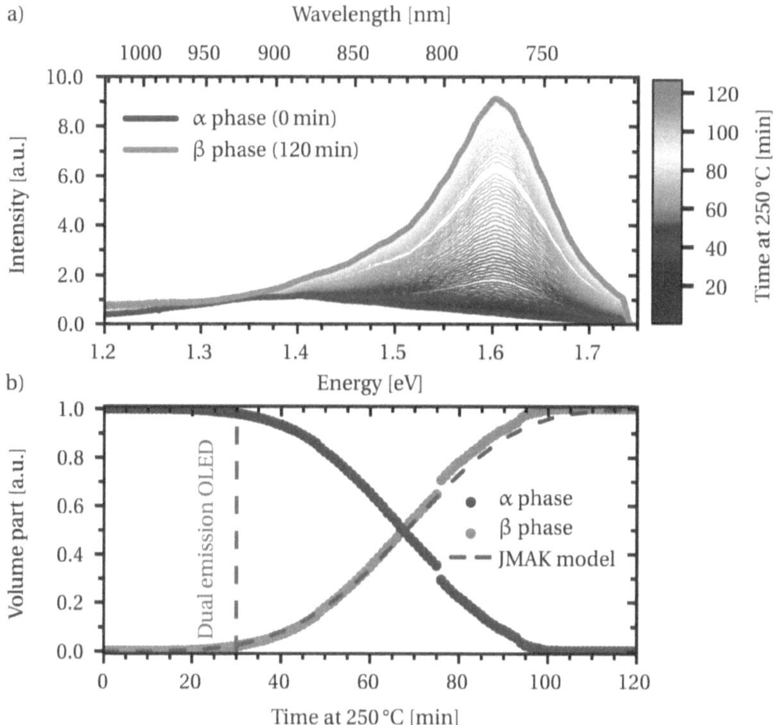

Figure 5.20.: a) Spectra evolving over the course of the α → β phase transition. Thick lines represent the neat spectra of the α (purple line) and β phase (red line). b) Volume part of the respective polymorphs extracted from spectral PL emission, including the fitted JMAK model for an isotropic 3D growth of the β volume fraction. Green dashed line marks the volume fraction at which the dual luminescent OLEDs are operated later on (*c.f.* section 5.5.2). Reproduced from [Ham+19], with the permission of AIP Publishing.

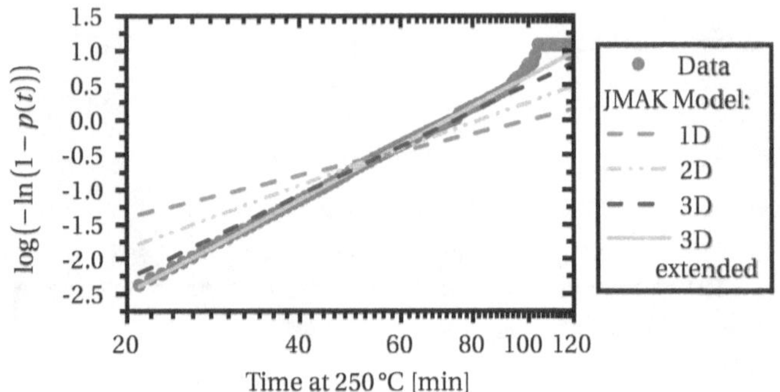

Figure 5.21.: Logarithmic plot to determine Avrami Exponent as slope of a linear function. 1D, 2D, and 3D JMAK model (dashed) as well an extended 3D JMAK model (solid cyan) has been used to reproduce the data.

In [Ham+19], the analysis of the phase transition kinetics ends with the disclosure of the polyhedral growth process, since a detailed examination of the mechanism of the phase transition was not the main goal of this publication. However, taking a closer look at figure 5.21, it becomes obvious that, even though being the best fit, the 3D JMAK model deviates slightly from the recorded data. The data follows a linear function, but the slope is slightly larger than the one of the 3D JMAK model. This deviation is caused by the assumption of a constant nucleation rate and allowing for a time dependent, transient nucleation rate leads to Avrami exponents larger than 4 [Pri+19][SM11]. Fitting equation (5.19) with n as free parameter yields an exceptionally good agreement with the data. The extracted Avrami exponent of $n = 4.46 \pm 0.02$ indicates a transient behavior of the nucleation rate, *i.e.* after the sample reached the annealing temperature the nucleation rate increases with time to its steady state value. The value is, however, very close to the Avrami exponent for a polyhedral growth with constant nucleation which justifies the assumption used in [Ham+19]. Sinha and Mandal [SM11] further derived that the Avrami exponent is not constant over time, especially at the beginning of the transition process. An even deeper analysis is, nevertheless, out of scope of this work. The analysis shows that the ZnPc phase transition can be very well explained by the JMAK model based on a polyhedral growth with homogeneous but slightly time dependent nucleation rate.

5.5.2. Dual Luminescent OLEDs

New concepts for thin, functional light emitting devices in the near infrared (NIR) are under intense discussion [Ker+15], proposing exciting possibilities to combine plasmonic coupling to nano antennas for spectrally and directionally selective emission from metal-organic hybrid devices. The implementation of metallic nanoparticles in an OLED for enhancing the device's quantum efficiency has been reported [Xia+12], and the successful enhancement of the α ZnPc excimer emission by coupling to plasmonic nanostructures has been demonstrated [KP17]. These promising findings motivated the exploitation of the spectral tunability of ZnPc thin films by means of controlling their phase transition to produce devices with a stable multi luminescent emission. The resulting devices comprise a single active layer, thus avoiding traps and emissive CT states at molecular heterojunctions, making electronic matching of different succeeding molecular interfaces obsolete. Last but not least, the single active layer simplifies the preparation process.

Three different types of OLED devices were prepared on ITO/glass substrates starting with a vacuum sublimed 30 nm thick ZnPc layer, followed by a BPhen capping layer acting as exciton blocking layer [Ste+12] and metal atom penetration barrier upon top contact deposition [Sch+05]. The Ca/Al top cathode is deposited through a shadow mask determining the dimensions of the two series of four quadratic devices to 3 mm × 3 mm and 5 mm × 5 mm on each substrate. A series of α phase OLEDs was produced using the ZnPc layer as deposited with a 10 nm thick BPhen capping layer resulting in an ITO/ZnPc (30 nm)/BPhen 10 nm/Ca (10 nm)/Al (150 nm) stack architecture. The β phase OLEDs were prepared by inducing the $\alpha \rightarrow \beta$ phase transition by annealing the ITO/ZnPc sample for one hour at 300 °C in a gas flow furnace. The successful phase transition on similar treated reference samples was confirmed by XRD measurements (*c.f.* section 5.3.2). Mixed film devices were prepared by annealing the ITO/ZnPc substrates on a hot plate at 250 °C. Three successive annealing steps of 60 s, 90 s, and 20 s were conducted while screening the spectrally resolved PL after each step to interrupt the phase transformation at the right time to achieve a balanced ratio of α and β phase emission. As indicated by the vertical dashed green line in figure 5.20 b) only a small fraction ($p \approx 0.02$) of β phase crystallites is needed for the balanced emission, due to the much higher quantum efficiency of the underlying Frenkel exciton emission. The enhanced thermal coupling of the ITO/ZnPc substrate to the hot plate compared to the gas flow furnace leads to a much faster and hence, difficult to control, phase transformation. For the neat β phase and mixed phase OLED devices, the BPhen capping layer thickness was increased to 30 nm to prevent pin hole induced shunts across the device as ITO is know to recrystallize above 250 °C [Yan+06][Rao+07] which could lead to newly formed ITO crystallites penetrating the subsequent organic layers.

The electrical device characteristics and the integrated EL intensities of the OLEDs normalized to the emission maximum are presented as a function of the applied electric field in figure 5.22 (left side). Here, the electric field turns out to be a more suitable quantity to compare the three device types as the varying BPhen layer thickness has to be taken into account. In figure 5.22 a), b) and c) the current density and the EL instensity are shown for an exemplary α phase, mixed phase, and β phase device, respectively. For all three devices, an overall diode like current density-voltage relation is found. For the thermally treated devices, the current density at low electric field strengths is much higher than for the untreated α phase diode. This is attributed to residual partial shortening caused by ITO crystals penetrating the subsequent organic layers after thermally induced recrystallization, despite the thicker BPhen layer is thought to compensate for most of the potential shunts. The basic functionality of the OLED, however, is not impeded by the occurring shunt resistances, as the EL onset is found at $2.0 \cdot 10^5 \, \mathrm{Vcm^{-1}}$ consistently for all three devices, indicated by the dashed blue line in figure 5.22 a)-c). Furthermore, for the annealed devices, the current density at high electric fields is significantly lower than for the untreated device. As the conductivity of the β phase is considerably lower than that of the α phase [SO86] and as a change of ITO stoichiometry upon annealing [Kim+05] lead to a lower anode conductivity, the overall lower current density can be attributed to thermally induced changes in conductivity upon heating.

In figure 5.22 d), e) and f) the spectrally resolved EL is shown for increasing field strengths $\vec{E}$ for the α phase, the mixed phase, and the β phase OLED, respectively. For the α and the β phase the observed EL emission corresponds to the photo excited emission, suggesting the same emission mechanism. A comparison between the thin film PL (c.f. figure) 5.9 c) and the EL of the neat b phase, not reported in [Ham+19], shows a slight red shift as well as a higher intensity in the subsequent vibronic shoulder at 1.5 eV. This suggests an additional intensity contribution by back reflection at the metal cathode which is subject to higher self absorption at the high energy side of the 0-0 transition (*c.f.* section 5.4.1) and hence, enhances the vibronic shoulder more than the 0-0 transition. The positions of the emission maxima of the neat crystal phases are indicated by dashed lines. The mixed phase emission in figure 5.22 e) clearly shows the presence of both emissions contributions confirming the successful preparation of a dual luminescent OLED from just a single active material.

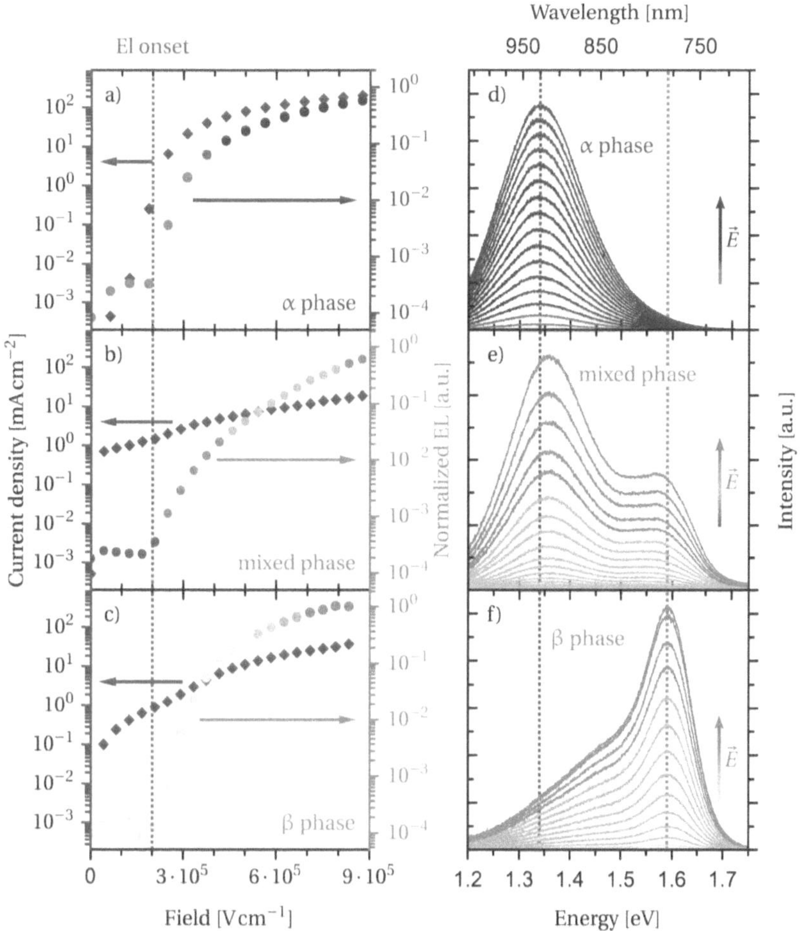

Figure 5.22.: Electrical device characteristics (left y-axis) and normalized integrated EL intensity (right y-axis) for α a), mixed b), and β phase c) ZnPc OLEDs. The matching EL onset is indicated by the dashed blue line. In d), e) and f) the spectrally resolved EL with increasing electrical field strengths $\vec{E}$ (indicated by arrow) for the corresponding devices is shown. The α and β phase emission maxima are marked by dashed lines. Reproduced from [Ham+19], with the permission of AIP Publishing. Arrows to indicate increase with electrical field strength in d), e), and f) have not been part of the original graphic and have been added for clarity.

To quantify the spectral composition in more detail, the ratio between the two distinct intensity maxima in the emission attributed to the ZnPc α and β phase, defined as

$$R = \frac{I\left(\lambda_{\text{max}, \beta}\right)}{I\left(\lambda_{\text{max}, \alpha}\right)} \tag{5.20}$$

can be utilized. It has to be noted, that the R as a spectral intensity ratio is not independent of the abscissa, due to the non-constant Jacobi transformation applied for transferring the spectra from the wavelength to the energy domain and, *vice versa*, as discussed in section 3.4.1. The R values discussed in the following are obtained from the wavelength dependent spectral data to minimize the effect of an artificial noise enhancement at low energies caused by the $1/E^2$ transformation and impeding a distinct estimation at low intensities. In figure 5.23 a) the R values are plotted as function of the current density (orange dots, upper abscissa). After reaching a sufficiently high intensity above $4\,\text{mA\,cm}^{-2}$, the unambiguous determination of the R values is possible. While the mean ratio is found at around 0.6 ± 0.1 (shaded pink area), a slight increase with increasing current density can be observed. This indicates an anisotropic distribution of β crystallites within the ZnPc layer resulting from asymmetric heating at the substrate side to induce the structural phase transformation leading to a strong temperature gradient within the ZnPc layer. This yields a higher fraction of β crystallites close to the ITO substrate. The fraction of β crystallites decreases with increasing distance from the ITO/ZnPc interface. Considering an asymmetric injection barrier for charge carriers, holes will be present within the device before electrons. Therefore, at low current densities, the recombination zone is found near the electron injecting top contacts and shifts towards the ITO contact with increasing charge carrier density, *i.e.* increasing electric field strength. This, in return, causes a higher contribution of the β phase to the observed EL emission for higher current densities. Figure 5.23 b) illustrates this scenario by two schematic devices indicating the position of the recombination zone for current densities corresponding to the limits of the upper abscissa in a). At high current densities the recombination zone resides in a region with an enhanced volume fraction of β phase crystallites, while for low current densities the emission originates from the region close to the top contact, *i.e.* of smaller β phase volume part.

For a possible implementation it is of utter importance that the spectral ratio of a multi luminescent device is stable over time. To exclude a possible ongoing $\alpha \rightarrow \beta$ transformation driven, *e.g.* by minimizing the surface tension at α/β phase grain boundaries or thermally induced local transitions by Joule heating, a long term study has been conducted over the course of $75\,\text{h}$. For this, an external voltage of $4.5\,\text{V}$, corresponding to a field strength of $7.5 \cdot 10^5\,\text{V\,cm}^{-1}$, has been applied to the device and

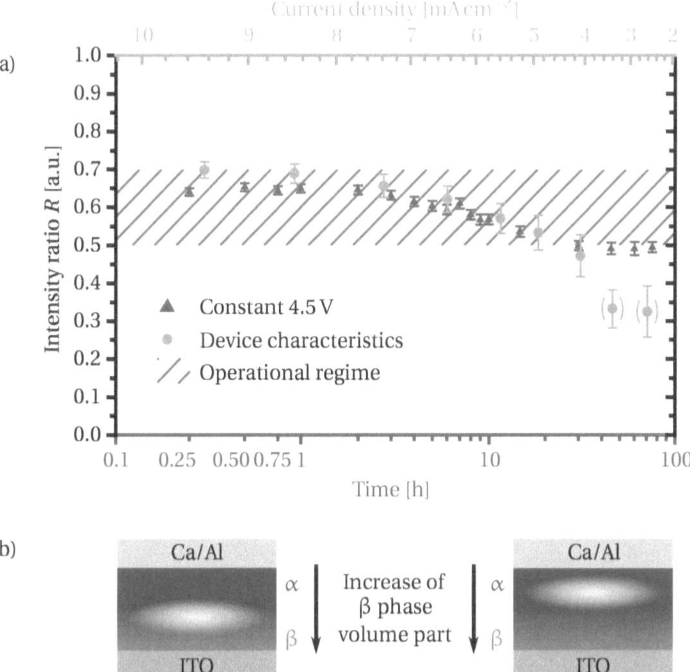

Figure 5.23.: a) Intensity ratio R over time at a constant applied voltage of 4.5 V (purple triangles) and as a function of current density (orange circles, upper abscissa) extracted from the ZnPc mixed phase OLED characteristics. The upper abscissa represents the decreasing current density at a given time as the Ca top contact degrades by oxidation. b) Schematics of the diodes indicating the shift of the recombination zone from the a well balanced region towards a region with α phase excess with decreasing current density. Reproduced from [Ham+19], with the permission of AIP Publishing.

the current density as well as the spectrally resolved emission have been recorded over time. The current density presented in the upper abscissa in figure 5.23 a) directly relates to the operating time presented as lower abscissa. The R values are plotted as function of operation time as purple triangles. During the first two hours of operation, the emission ratio is stable at about $R = 0.65$. This regime is followed by a steady decline over the next 10 h before leveling at roughly 0.5. As indicated by the upper abscissa, the decline in the emission ratio coincides with a drop in current density. In fact, the emission ratio extracted from the device characteristics and associated with a shift of the recombination zone closely resembles the observed shift in emission ratio with operation time. This suggests that the slight change in emission behavior is not associated to a change of the ratio of the ZnPc polymorphs in the active layer, but rather with a shift of the recombination zone caused by the decline in the current density. The observed decline in current density can be ascribed to the gradual oxidation of the Ca top contact, which can be prevented by a suitable encapsulation in a future device optimization which was not part of the present study.

The above presented investigation demonstrates the feasibility of a multi luminescent OLED using just a single active layer. The understanding of the individual radiative and non-radiative mechanisms leading to the enhanced quantum efficiency of the β phase emission in combination with the control of the phase transition kinetics results in the design of a well balanced dual luminescent OLED based on a single ZnPc emission layer.

5.6. Electronic Phase Transition: Exciton Delocalization and Superradiance

So far, the photophysics of the ZnPc α phase emission has been related to an excimer emission, whereas the β phase polymorph emission, especially at low temperatures, has not been discussed to the same extend. The ZnPc β phase polymorph in the form of a single crystal shows an interesting behavior with temperature. The emission spectra as well as its overall intensity are the result of an interplay between molecular packing and mono molecular processes. In section 5.4.2 two temperature regimes of the integrated emission intensity dominated by different photophysical processes have been presented. While at high temperatures a temperature activated emission behavior associated with a mono molecular radiative decay is found, the low temperature regime is in excellent agreement with a thermally induced localization of initially delocalized crystal excitons in a J-aggregate. The room temperature crystal structure suggests a J-type molecular coupling in agreement with the thin film absorption (*c.f.* section 5.4.1) and the narrow emission line of the single crystal below 100 K, associated with the 0-0 transition, corroborates this interpretation. Near room temperature, as discussed in section 5.4.4, the emission spectrum is consistent with a Frenkel exciton emission accompanied by vibronic progression. Upon cooling the emission changes to an H-aggregate like behavior accompanied by a decrease in overall intensity. In figure 5.10 b) it can be seen that the emission nearly vanishes before the narrow emission line appears and the overall intensity, depicted in figure 5.11 b), rises again. Such a sudden change in physical properties with temperature is usually associated with a *phase change* and often accompanied by a change in the underlying crystal structure. Interestingly, for phthalocyanines no such structural phase transition is known for low temperatures. Furthermore, preliminary results of temperature dependent XRD measurements down to 77 K performed within a collaborative project by Mehmet Özcan and Bert Nickel, Ludwig-Maximilians University München, show no indication of a structural phase transition. However, the spatially anisotropic thermal contraction well-known for metal phthalocyanines has been observed. The continuous change of the crystal lattice, *i.e.* the expansion and contraction of the unit cell, leads to dramatic changes in the optical properties of the ZnPc β phase crystal, which will hence be referred to as an "*electronic phase transition*" to emphasize the dramatic modification in the optical properties without an accompanying structural phase transition.

In the following section the low temperature optical state of the ZnPc β phase crystal will be discussed. The photoluminescence spectra are interpreted based on the Spano exciton polaron theory indicating exciton delocalization as the driving force for the observed emission characteristics. The transition dipole moments are examined by means of polarization dependent photoluminescence studies. Finally, a qualitative model based on the room temperature crystal structure and dipole-dipole interaction is derived to consistently explain the superradient behavior as well as the emission polarization.

Temperature and emission polarization dependent PL studies have been performed as described in section 3.4.2 under circular polarized 685 nm cw laser excitation. The sample and the Si diode for temperature control were fixed with silver paint on a copper plate and mounted into the cryostate as described in section 5.4.2. The long axis of the ZnPc crystal corresponds to the 90° emission polarization. The polarization was probed between 0° to 360° in 20° steps. Spectra were recorded with the 300 lines mm^{-1} optical grating at a central wavelength of 825 nm and with an acquisition time of 20 s.

The following data has been obtained together with Larissa Lazarov, Julius-Maximilians University Würzburg, as part of her final book. The qualitative model, presented in section 5.6.3, has been developed in close cooperation during her book.

5.6.1. Exciton-Polaron Coupling

At the beginning, the low temperature PL spectra will be analyzed. As discussed briefly in section 5.4.2, below 100 K the PL spectra differ from those of a broad Frenkel exciton emission and a narrow emission line containing most of the overall intensity accompanied by well resolved vibronic satellites constitutes the emission spectrum. To examine this spectral composition further the 0° polarization emission, corresponding to the highest detected intensity (*c.f.* figures 5.25 a) and 5.26), is investigated.

In figure 5.24 a) the 5 K emission spectrum is shown over energy shift with respect to the detected 0-0 transition at 1.624 eV. The upper x-axis corresponds to the overall emission energy. Clearly, the main transition peak shows a splitting into two distinct peaks whose origin could not be elucidated, so far. Both peaks show a similar polarization dependency indicating that the involved transition dipole moments are oriented parallel to each other, thus, excluding a Davydov splitting. However, the high energy component seems to fully vanish at 90° polarization, whereas a low intensity of the low energy component is still detectable (*c.f.* appendix D.3). The energetic splitting is in the order of 7 meV and, even though vibrational out of plane modes have been predicted for ZnPc in this energy range [Pal+95][TDE00], they have not been reported to play a role in Franck-Condon absorption spectra [The+15][Fen+20]. Furthermore, in the solvent emission spectra (*c.f.* section 5.2.2) no contribution of a low energy vibrational mode was found. In case of a two dimensional exciton delocalization Eisfeld *et al.* [Eis+17] predict exciton eigenstates separated by a similar energetic spacing contributing to the overall emission upon thermal occupation. Given these results and predictions, it is assumed that both peaks are related to the 0-0 transition in the Spano superradiance picture [SY11].

Bedides the pronounced 0-0 transition numerous peaks, all associated with vibronic transitions, are observed. At 25 meV and 55 meV two previously not resolved vibrational modes, ζ and ξ, are found. Overall, the various vibronic transitions add up in intensity to an almost constant background occasionally protruded by a distinct emission peak. At about 80 meV a sharp peak is visible and the energetic position discloses the relation to the vibrational α mode, as found for the high temperature β phase emission and the single molecule emission in DMSO (*c.f.* sections 5.2.2 and 5.4.4). At the energetic positions of the other two previously employed vibrational modes β (140 meV) and γ (180 meV) which shape the emission spectra at higher temperatures, no unambiguous assignment is possible. According to equation (2.60), the intensity ratio of the 0-0 and 0-1 transitions constitute a measure for the superradiant enhancement η_{SR} and is related to the number of coherently excited molecules N_{coh} via the Huang-Rhys factor S:

$$\eta_{SR} = \frac{I_{0-0}}{I_{0-1}} = \frac{N_{coh}}{S}. \tag{5.21}$$

The quantity S is usually considered an "effective Huang Rhys parameter" averaging over all vibrational modes and being experimentally determined [Eis+17]. The energetic broadening of the vibronic background and the unknown origin of the 0-0 peak splitting render a decomposition of the full spectrum difficult if not impossible. While previous publications [Eis+17][Mül+13] fit the 0-0 transition with a suitable line shape function and integrate the intensity of all vibronic progressions, including higher vibronic transitions, resulting in a lower estimate for the superradiance enhancement parameter η_{SR}^-, here, the dominant α mode can be used to give an upper estimate of η_{SR}^α as well.

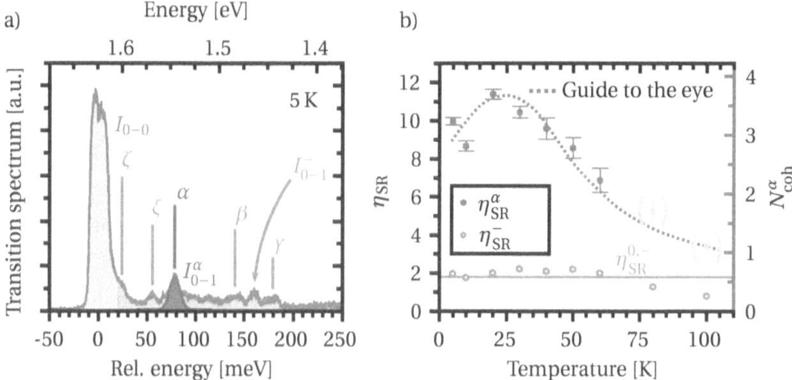

Figure 5.24.: a) ZnPc crystal PL spectrum at 5 K at 0° polarization. Vibronic transitions are labeled accordingly. Integrated intensities I_{0-0} (red), I_{0-1}^α (purple) and, I_{0-1}^- (grey) are indicated as shaded areas. b) Enhancement factors η_{SR}^α (blue) and η_{SR}^- (grey) over temperature. The number of coherently excited monomers N_{coh}^α from the alpha phase estimation is given by the right y-axis. The upper limit for single molecule limit $\eta_{SR}^{0,-}$ of η_{SR}^- is indicated by the horizontal grey line.

The intensity of the 0-0 transition is estimated by integrating transition spectrum over the relative energy difference in figure 5.24 a) from $-25\,\text{meV}$ to $20\,\text{meV}$. The upper boundary is chosen in such a way excluding contributions by the ζ transition while simultaneously not underestimating the 0-0 transition. The lower boundary of η_{SR} is obtained by integrating the intensity of the vibronic tail over the of the relative energy from 20 meV to 200 meV (I^-_{0-1}), intentionally including various mixed and higher vibronic transitions from the α, ζ, and ξ mode. For an upper limit, only the 0-1 α transition has to be considered. The intensity I^α_{0-1} is estimated by a Gaussian function fitted to this peak in a narrow energy range while keeping the baseline fixed. An exemplary fit is shown in figure 5.24 a) (purple). For 80 K and 100 K the α mode feature is below the detection threshold and, thus, has been estimated by integration between 55 meV and 105 meV. In figure 5.24 b) the upper (blue data points) I^α_{0-1} and lower (grey data points) estimation η^-_{SR} of the superradiance enhancement factor η_{SR} is plotted as function of temperature. For a single molecule, assuming a transition from the vibrational ground state of the excited state justified for low temperatures, equation (5.21) can be evaluated analytically yielding the single molecule limit η^0_{SR} of η_{SR}. For a set of vibronic modes J with the respective Huang-Rhys parameters S_j for $j \in J$, η^0_{SR} is given by calculation of the respective Franck-Condon factors defined by equation (2.40):

$$\eta^0_{\text{SR}} = \frac{I_{0-0}}{I_{0-1}} = \frac{\prod\limits_{j \in J} e^{-S_j}}{\sum\limits_{j \in J} S_j \prod\limits_{j \in J} e^{-S_j}} = \frac{1}{\sum\limits_{j \in J} S_j}. \tag{5.22}$$

From the analysis of the high temperature emission in section 5.4.4 it is known that the Huang-Rhys parameters of the different vibrational modes are heavily influenced by the surrounding crystal lattice and hence, are likely to be temperature dependent. As reference points the values at 320 K are chosen. This choice is motivated by the assumption that the low temperature opitcal emission is induced by a molecular J-aggregate which is confirmed by the room temperature crystals structure (*c.f.* section 5.4.1). Based on the observed emission spectra, the molecular packing of the J-aggregate is most likely present also at lower temperatures. Furthermore, at 320 K S_α as well as $S_\alpha + S_\beta + S_\gamma$ have the smallest values and hence constitute an upper limit for η^0_{SR}.

By the estimated lower limit η^-_{SR} of the surradiance enhancement the single molecule limit can be assessed upwards using equation (5.22) with $S_\alpha + S_\beta + S_\gamma$ as denominator, hence, neglecting contributions by other vibrational modes. this approach yields an upper limit to $\eta^{0,-}_{\text{SR}} \leq 1.770 \pm 0.029$ of the single molecule enhancement factor which is indicated by the horizontal grey line in figure 5.24 b), suggesting that for temperatures below 80 K the single molecule emission does not define a sufficient measure for the observed η_{SR} values. For the upper estimation of the superradiant enhancement η^α_{SR} the number of coherently excited molecules N^α_{coh} calculated by equation (5.21) with $S_\alpha = 0.325$ is given on the right y-axis. From 5 K to 20 K a slight increase in η^α_{SR} and, hence, in N^α_{coh}, is observed. With increasing temperature, scattering with phonons leads to occupation of states at the bottom of the exciton band with a quasi-momentum $k > 0$. For these states a direct electronic transition to the vibrational ground state is momentum forbidden, thus decreasing the intensity of the 0-0 transition. In contrast, the transition to the first vibrationally excited state of the electronic ground state and, hence the intensity of the 0-1 transition, is not impeded by this restriction due to the generation of a vibrationally excited state ensuring conservation of momentum. Hence, the number of coherently excited monomers decreases to 1 at 100 K where the single molecule limit is reached and a localized Frenkel exciton dominates the emission process.

The results above show that the estimation of the superradiant enhancement η_{SR} surpasses the single molecule limit in case of the upper as well as of the lower estimation indicating a delocalization of the excitons in the crystal lattice. As the Huang-Rhys factors are temperature dependent and as not all vibrational modes can be examined the estimation of the number of coherently excited molecules according to the Spano method in equation (5.21) has to be handled with care. A more reliable indicator for the delocalization of excitons is the exceeding of the single molecule limit η_{SR}^0 for both estimations η_{SR}^- and η_{SR}^α of the superradiant enhancement factor. Furthermore, with increasing temperature decoherence limits the exciton delocalization to the point where the emission is dominated by a localized Frenkel exciton.

5.6.2. Temperature Dependent Emission Polarization

While in the previous section the composition of the emission spectrum at one particular polarization was examined, leading to a conclusive picture of exciton delocalization, the following section focuses on the polarization dependency of the emission to relate the radiative process to the spatial anisotropy along the crystallographic directions.

To determine the degree of polarization, the normalized integrated intensity as a function of the polarization angle $I(\rho)$ was fitted by

$$I(\rho) = P\cos^2\left(\rho + \phi\right) + I_0 \tag{5.23}$$

where ϕ is a phase offset with respect to the 0° position of the polarizer. The amplitude P and the constant offset I_0 determine the degree of polarization. For unpolarized light $P = 0$ leaving $I(\rho) = I_0$ independent of the polarization angle. For $I_0 = 0$ the emitted light is linearly polarized along the ϕ direction. A representative fit of equation (5.23) to the data set for 5 K is shown in figure 5.25 a) together with the three fit parameters P, ϕ and, I_0.

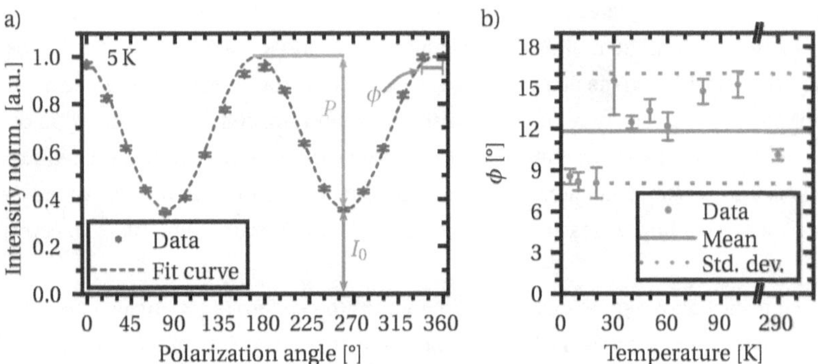

Figure 5.25.: a) Representative polarization fit of the normalized intensity at 5 K. The impact of the parameters P, I_0, and ϕ on the fit function is indicated in the graph. b) Phase offset ϕ obtained by fits as function of temperature together with mean value (solid line) and plus/minus one standard deviation (dotted lines).

The resulting fit curves between 5 K and 100 K together with a room temperature reference are shown as a polar plot of the polarization angle in figure 5.26. All curves are normalized to their maximum emission. The crystal (exemplary photograph shown as background) was aligned along the 90° respectively 270° polarization axis. At low temperatures, the emitted luminescence is highly polarized along the 170° respectively 350° direction, *i.e.* along the direction of the phase offset ϕ. However, with increasing temperature the polarization is weakened. At room temperature, even though a preferred direction still exists, the emission is mostly unpolarized. The polarization direction does not change with temperature. The phase offset ϕ obtained from the polarization fit is plotted in figure 5.25 b) as function of temperature. Despite some minor variation, all ϕ values lie at around 11.8° within one standard deviation of ±2.9°, keeping the polarization angle fixed. As discussed in the following section 5.6.3, this most likely indicates that the polarization direction is perpendicular to the crystal's long axis. The small phase offset ϕ with respect to the 0° respectively 180° polarization axis is associated with a slight misalignment of the crystal in the optical measurement which is amplified by the angular sensitivity of the half wave plate.

To quantify the loss of polarization with temperature, the *linear degree of polarization LDOP* is used, which is defined according to [Dem17] as

$$LDOP = \frac{I_\parallel - I_\perp}{I_\parallel + I_\perp} \tag{5.24}$$

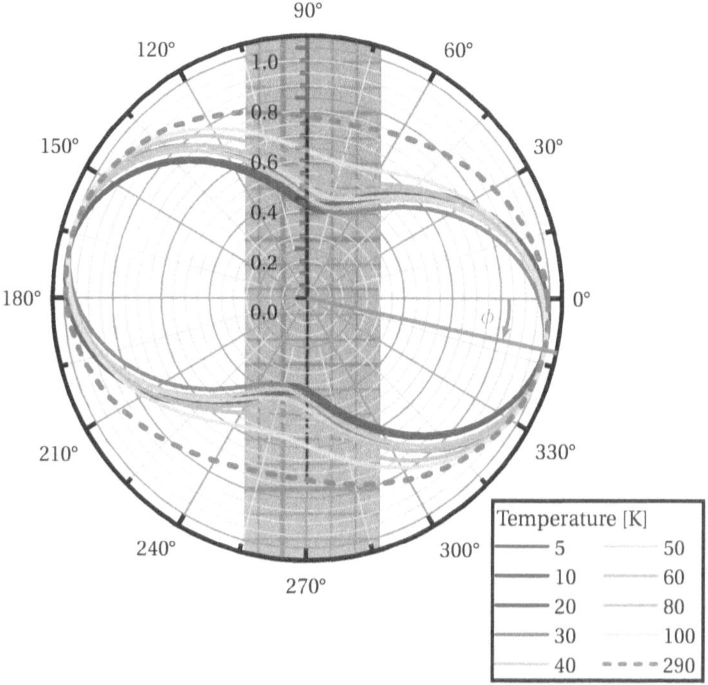

Figure 5.26.: a) Polar plot of the fitted normalized emission intensity as function of the polarization angle for several temperatures. The mean value of phase offset ϕ indicates a slight tilting with respect to the coordinate system of the sample illustrated by the photograph of a ZnPc crystal on graph paper along the 90° respectively 270° polarization direction behind the polar plot.

where $I_\parallel$ is the intensity at a polarization parallel to a particular direction and $I_\perp$ is the intensity at a polarization perpendicular to this direction. The *LDOP* for an emission described by equation (5.23) along the ϕ direction is given by

$$LDOP = \frac{P}{P + 2I_0} \quad .$$

(5.25)

The obtained fit parameters P and I_0, as well as the calculated *LDOP* are plotted as a function of temperature in figure 5.27 a), b) and, c), respectively. As can be seen, P and I_0 show an anti correlated behavior with temperature. While for temperatures below 40 K P and I_0 are constant, a change in both properties occurs with rising temperature. This translates directly to the *LDOP*. For values below 40 K the polarization degree appears to be only slightly below *LDOP* = 0.5 which corresponds to a strong polarization along the ϕ direction, *i.e.* to an emission ratio of parallel to perpendicular polarized intensity of $I_\parallel : I_\perp \approx 3 : 1$. From there, the emission becomes less polarized with increasing temperature until, at 100 K, an *LDOP* value of approx. 0.3 is reached which corresponds to an emission ratio of $I_\parallel : I_\perp = 2 : 1$.

The *LDOP* behavior with temperature is similar to the one of the superradiance enhancement factor $\eta_{\mathrm{SR}}^{\alpha}$. Figure 5.27 d) shows a scatter plot of data pairs $\left(\eta_{\mathrm{SR}}^{\alpha}(T), LDOP(T)\right)$, thereby correlating *LDOP* values with the respective $\eta_{\mathrm{SR}}^{\alpha}$ values of the same temperature. A correlation analysis as well as a linear fit, also shown with its 95 % confidence band in figure 5.27 d), yield a Pearson r value of $r = 0.98721$ indicating a strong linear correlation of both quantities. This confirms that the exciton delocalization and its thermally induced localization are related to the loss of polarization of the ZnPc crystal luminescence. The following the microscopic origin of this correlation will be analyzed showing that both phenomena originate from the same photophysical process.

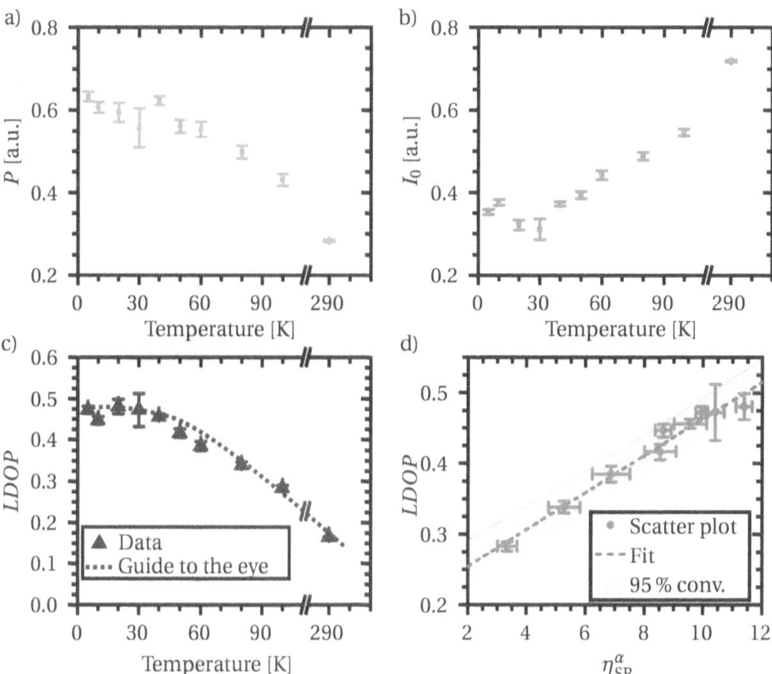

Figure 5.27.: Polarization fit parameters a) P and b) I_0 as function of temperature. c) Linear degree of polarization (LDOP) as function of temperature. the Guide to the eye indicates a depolarization of the emitted luminescence with increasing temperature. d) Scatter plot of $\left(\eta_{\mathrm{SR}}^{\alpha}(T), LDOP(T)\right)$ data pairs to examine the correlation between exciton delocalization and the luminescence polarization. The Linear fit yields a Pearson r value of 0.987 21.

5.6.3. Inter Chain Coupling via Dipole-Dipole Interaction – A Qualitative Model

To examine the correlation between the exciton delocalization and the emission polarization, the excitonic coupling in the $(\bar{1}01)$ lattice plane, which corresponds to the examined ZnPc β phase crystal surface (*c.f.* section 5.3.1), is investigated based on a dimer dipole-dipole coupling. In figure 5.28 a), the molecular packing in the $(\bar{1}01)$ plane is depicted together with the projected unit cell indicated by the black dashed lines based on the crystal structure reported in [Luc+20]. The plane comprises stacks of parallel molecules along the crystallographic $\vec{b}$ direction, resulting in a herringbone packing pattern. For a given molecule in the plane there exist eight molecules in the immediate vicinity as shown for the highlighted molecular position 0 surrounded by the cyan frame in figure 5.28 a). With regard to these eight sites, only the molecules at position 1, 2, and 3 lead to a nonequivalent dipole-dipole coupling with the molecule at the central position. For the other positions the crystal symmetry leads to an equivalent molecular orientation and hence dipole-dipole coupling which is indicated by the respective label marked with an apostrophe. The molecules constituting one molecular stack (position 1) are equivalent due to their translation symmetry along the $\vec{b}$ direction. The inversion symmetry of the unit cell leads to the equivalency of the four molecules sitting in the corners of the unit cell (position 3) as are the next nearest neighbors in the neighboring stacks (position 2).

The ZnPc's molecular transition dipole moment lies within the molecular plane as shown by the orange vector $\vec{M}_0$ in figure 5.28 b) [Zha+16] and points from the central Zn atom to a C atom on the inner C−N alternating ring as described in section 5.4.1. Due to the molecular symmetry a second transition dipole moment is to be expected perpendicular to $\vec{M}_0$ and hence, perpendicular to the $(\bar{1}01)$ crystal plane and not contributing to the observed emission. For this reason only the in-plane coupling is considered and, as an approximation, it is assumed that the dipole moment of relevance is embedded completely within the $(\bar{1}01)$ plane. This enables the defintion of a two dimensional intra molecular coordinate system shown in figure 5.28 b). The dipole-dipole interaction between the central molecule 0 and any given molecule i depends only on the position $\vec{r}_{0i}$ of molecule i and the two in-plane angles α and β. Expressing the inter molecular position and the two transition dipole moments $\vec{M}_0$ and $\vec{M}_i$ in the intra molecular coordinate system yields

$$\vec{r}_{0i} = r_{0i} \begin{pmatrix} \cos(\beta) \\ \sin(\beta) \end{pmatrix}, \quad \vec{M}_0 = M_0 \begin{pmatrix} 1 \\ 0 \end{pmatrix}, \quad \text{and} \quad \vec{M}_i = M_0 \begin{pmatrix} -\cos(\alpha - \beta) \\ \sin(\alpha - \beta) \end{pmatrix}. \tag{5.26}$$

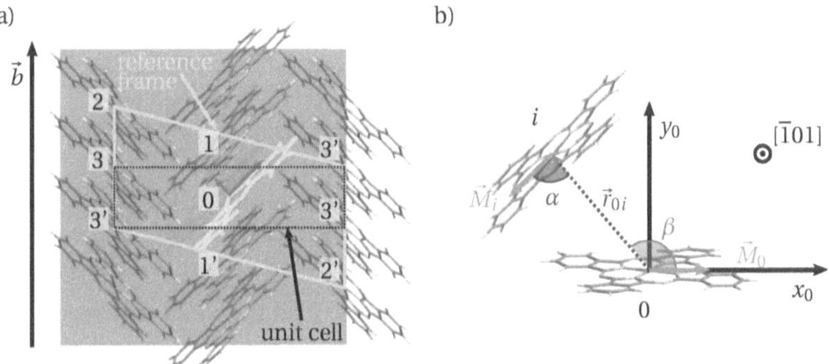

Figure 5.28.: a) Molecular packing in the ($\bar{1}$01) lattice plane together with the projected unit cell (dashed black frame). The Dipole-Dipole interaction to the nearest (position 1 and 3) and next nearest (position 2) neighbors of the central molecule at position 0 (included in the cyan frame and highlighted in yellow) is estimated. Equivalent positions originating by the crystal symmetry are marked with an apostrophe. The intra molecular coordinate system is indicated by grey arrows. b) The intra molecular coordinate system of the reference molecule at position 0 lies in the ($\bar{1}$01) lattice plane. The dipole-dipole interaction with a molecule at position i is defined by the inter molecular distance $\vec{r}_{0i}$ and the angles α and β. The in-plane molecular transition dipole moments of both molecules $\vec{M}_0$ and $\vec{M}_i$ are indicated in orange.

Based on the defintions in (5.26), the resonance energy E_{0i} of the dipole-dipole interaction, as given by equation (2.47), can be written as

$$E_{0i} = \frac{M_0^2}{4\pi\epsilon_0} \frac{1}{r_{0i}^3} \left(3\cos(\beta)\cos(\alpha) - \cos(\alpha - \beta) \right) \tag{5.27}$$

where M_0 is the magnitude of the ZnPc's molecular transition dipole moment. The resulting transition dipole moment is given by

$$\vec{M}_{0i} = \frac{M_0}{\sqrt{2}} \begin{pmatrix} 1 - \cos(\alpha - \beta) \\ \sin(\alpha - \beta) \end{pmatrix}. \tag{5.28}$$

As described in section 2.3.2, the dimer resonance interaction leads to an upper and lower energy state depending on the respective orientation of the two dipole moments. Dictated by Kasha's rule, only the energetically lower lying state contributes to the emission. For this reasons, each dipole-dipole calculation is performed for two configurations of the transition dipole moment $\vec{M}_i$ of molecule i , *i.e.* α and $\alpha + 180°$. The resulting transition dipoles are labeled by superscript l for the energetically lower lying state and by u for the upper one.

Molecule i	Distance r_{0i} [Å]	α [°]	β [°]
1	4.853	45.48	180-134.52
2	12.035	10.46	97.77
3	9.886	31.47	120.58

Table 5.7.: Coordinates extracted by Mercury 4.0 [Mac+20] from the ZnPc crystal structure reported in [Luc+20] and used for the dipole-dipole interaction calculations.

The coordinates for the calculations are extracted from the published crystal structure of Luc *et al.* [Luc+20] with Mercury 4.0 [Mac+20] and listed in table 5.7. The inter molecular distance was measured between the central Zn atoms and the angles where measured between the C_1–Zn conjunction of one molecule and the connection between its central Zn atom to the Zn atom of the adjacent molecule as shown in appendix D.4. Using the room temperature β phase crystal structure, as well as the first order dipole-dipole approximation, the results have to be handled with care and are interpreted more in a qualitative than quantitative fashion. The calculated energetic splittings $\Delta E_{0i} = 2E_{0i}$, indicating the interaction strengths as well as the magnitude of the resulting transition dipole moments $\vec{M}_{0i}^l$ and $\vec{M}_{0i}^u$ are listed in table 5.8 in units of $M_0^2/(4\pi\epsilon_0) \cdot 10^{-3}\,\text{Å}^{-3}$ and $M_0/\sqrt{2}$, respectively.

Molecular pair $0i$	$\Delta E_{0i}\left[\frac{M_0^2}{4\pi\epsilon_0}\cdot 10^{-3}\text{Å}^{-3}\right]$	$\lvert\vec{M}_{0i}^l\rvert$ [$M_0/\sqrt{2}$]	$\lvert\vec{M}_{0i}^u\rvert$ [$M_0/\sqrt{2}$]
01	8.309	2	0
02	0.511	1.381	1.447
03	2.727	1.403	1.425

Table 5.8.: Energetic splitting $\Delta E_{0i} = 2E_{0i}$ in units of $M_0^2/(4\pi\epsilon_0) \cdot 10^{-3}\,\text{Å}^{-3}$ as well as resulting transition dipole moment strengths in units of $M_0/\sqrt{2}$ for energetically lower and higher lying dimer state, $\vec{M}_{0i}^l$ and $\vec{M}_{0i}^u$, respectively.

The dipole-dipole calculation for the molecular pair 01 reveals the strongest interaction as well as a J-type coupling for which the transition strength is doubled for the low energy state, whereas the transition from the high energy state is dipole forbidden. While the qualitative J-aggregate coupling has been reported in section 5.4.1, the high interaction strength is a key result of the dipole-dipole calculations, suggesting the intra chain coupling along the crystallographic $\vec{b}$ direction, *i.e.* along the molecular stacks to be the dominant photophysical process. The excited state of the super molecular aggregate thus constitutes a crystal exciton delocalized along a linear chain as described in section 2.3.3. The coherent coupling of N molecules along the molecular chain yields an enhancement of the radiative transition rate by a factor N according to equation (2.58) and the resulting transition dipole moment $\vec{M}_J$ of the delocalized "chain exciton" is aligned parallel to $\vec{M}_0$ as the interacting molecules are cofacially packed. Thus, the J-type coupling along the crystallographic $\vec{b}$ direction constitutes a first insight into the microscopic photophysical effects leading to superradiant emission below 100 K. Considering excitons on two neighboring chains, their transition dipole moments are aligned almost perpendicular to each other as indicated in figure 5.29 a) by the green arrows. Thus, if the superradiance obtained at low temperatures would be driven solely by an intra chain coupling, the detected emission polarization should resemble a source of unpolarized luminescence in contraction to the observed results. This suggests the need of a more detailed picture to include a polarized emission at low temperatures. To explain the progression of the emission polarization with temperature the following scenario is suggested.

Below 100 K exciton delocalization along the molecular chains is favored leading to a superradiant emission. While at temperatures below 40 K the emission is highly polarized the degree of polarization decreases with temperature. At the high end of the temperature range, the excitation coupling occurs preferably along the stacking direction, indicated by the much higher resonance energy found for the 01 molecular dimer. Hence, the observed steady state emission is likely to exhibit a low degree of polarization due to the orientation of the resulting transition dipole moments with respect to each other. Upon further cooling the excited states on neighboring chains can couple which each other resulting in additional exciton delocalization within the plane. Approximating this inter chain coupling by a dipole-dipole interaction between two adjacent $\vec{M}_J$ transition dipole moments at distance of $r_C = \sqrt{(1/2 r_{01})^2 + r_{03}^2} = 10.179\text{Å}$, according to the crystal structure, yields two states with approximately the same transition dipole strength $|\vec{M}_J|$. In figure 5.29 a) the resulting transition dipole moments $\vec{M}_C^u$ and $\vec{M}_C^l$ for the inter chain coupling are indicated with respect to the molecular packing. Complementary, in b) the energetic splitting and the resulting transition dipole moments for the intra and inter molecular coupling are shown. In case of the inter chain coupling, the transition dipole moment of the energetically lower excited state is oriented perpendicular to the $\vec{b}$ axis of the crystal explaining the observed polarization of the low temperature photoluminescence.

The qualitative model above provides a first insight into the nature of the observed electronic phase transition. Although, according to the current information on the temperature dependent data no structural phase transition is assumed for the observed change in luminescence at around 100 K, the molecular packing nevertheless determines the photophysics. The molecular packing along the crystallographic $\vec{b}$ di-

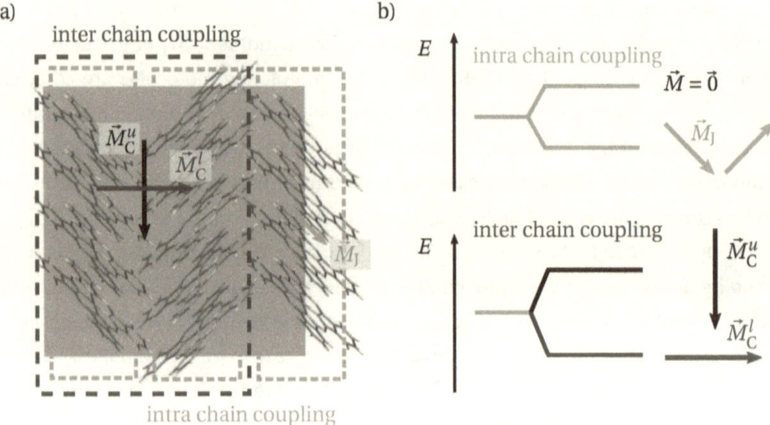

Figure 5.29.: Schematic illustration of intra and inter chain coupling a) with respect to the molecular packing and b) with respect to the energetic splitting.

rection promotes exciton delocalization yielding an enhancement of the 0-0 transition. The strong transition dipole moments on each molecular [010] chain can then couple with those on neighboring chains, and the near perpendicular molecular orientation of the herringbone packing yields an emission polarized perpendicular to the crystallographic $\vec{b}$ direction, *i.e.* to the long crystal axis. Hence, the above model provides a first idea on how the different effects are interrelated. The inter chain exciton dipole coupling as the necessary condition for the observed spectra is corroborated by preliminary quantum chemical calculation based on the room temperature crystal structure performed by Evgenii Titov, Julius-Maximilians University Würzburg, and have privately communicated to the author. These promising results together with the temperature dependent XRD analysis of the crystal structure constitutes a promising and versatile starting point for further investigation and potential application of the electronic low temperature phase of ZnPc single crystals.

6. Summary

This book focused on the influence of the underlying crystal structure and hence, of the mutual molecular orientation, on the excited states in ordered molecular aggregates. For this purpose, two model systems have been investigated. In the prototypical donor-acceptor complex pentacene-perfluoropentacene (PEN-PFP) the optical accessibility of the charge transfer state and the possibility to fabricate highly defined interfaces by means of single crystal templates enabled a deep understanding of the spatial anisotropy of the charge transfer state formation. Transferring the obtained insights to the design of prototypical donor-acceptor devices, the importance of interface control to minimize the occurrence of charge transfer traps and thereby, to improve the device performance, could be demonstrated. The use of zinc phthalocyanine (ZnPc) allowed for the examination of the influence of molecular packing on the excited electronic states without a change in molecular species by virtue of its inherent polymorphism. Combining structural investigations, optical absorption and emission spectroscopy, as well as Franck-Condon modeling of emission spectra revealed the nature of the optical excited state emission in relation to the structural α and β phase over a wide temperature range from 4 K to 300 K. As a results, the phase transition kinetics of the first order $\alpha \rightarrow \beta$ phase transition were characterized in depth and applied to the fabrication of prototypical dual luminescent OLEDs.

Pentacene Crystal Structure and PFP/PEN Interface Morphology

As a precondition for the spectroscopic investigation and interpretation of charge transfer formation, especially, for the head-to-tail orientation of the molecules, a thorough structural and morphological characterization of the neat substrate PEN crystals as well as of the PFP/PEN crystal interface were performed. X-ray diffraction (XRD) measurements on neat PEN single crystals confirmed the expected bulk crystal phase for the sublimation grown crystals by their 14.1 Å out-of-plane (001) spacing [Mat+01]. However, the diffraction patterns of several samples reliably indicate the coexistence of a second structural phase characterized by a 14 3 Å out-of-plane lattice distance This is most likely caused by a residual crystalline polymorph formed during crystal growth. Atomic force microscopy (AFM) performed on the neat crystal surface reveals large terraces separated by monomolecular steps, confirming the high quality of the samples and the upright orientation of the PEN molecules at the (001) crystal surface. Upon deposition of a sub monolayer PFP on a PEN (001) crystal surface additional phase sensitive AFM measurements verified the upright orientation of PFP molecules right at the PFP/PEN (001) crystal interface exhibiting a head-to-tail molecular orientation. Further XRD studies of PFP thin films grown on PEN single crystals strongly suggest that the PFP film is stabilized in its thin film phase [Sal+08a].

Charge Transfer Formation in Archetypical Donor-Acceptor Interfaces

Based on this structural data, the charge transfer state formation in PEN-PFP was investigated for two prototypical molecular orientations, the cofacial and the head-to-tail stacking. The cofacial stacking was realized in PFP doped PEN thin films and bulk heterojunction diodes by coevaporation at a PEN:PFP molecular ratio of 12:1 and 3:1, respectively. Using UV/Vis absorption spectroscopy an additional absorption band at 1.57 eV, absent in the neat materials, is found, confirming the formation of a charge transfer state in accordance with literature [Bro+11]. The head-to-tail orientation of the donor-acceptor system was investigated using PEN (001) single crystals as substrates for the subsequent PFP growth, yielding a well defined standing stack molecular interface. The interfaces were characterized by temperature dependent photoluminescence studies, as well as by differential reflectance spectroscopy (DRS), disclosing, for the first time, the absence of charge transfer state formation in the molecular head-to-tail configuration. Both interface geometries were implemented in prototypical bilayer (head-to-tail orientation) and bulk heterojunction (cofacial orientation) diodes. Their optoelectronic characterization revealed an electroluminescence centered at 1.4 eV and associated with cofacial charge transfer emission [Ang+12]. However, in the bulk heterojunction devices, the much higher charge transfer state density leads to an overall lower current density upon ambipolar charge injection as the charge transfer states act as charge carrier traps. These results highlight the importance of controlling the molecular orientation at donor-acceptor interfaces to maximize their efficiency in heterojunction devices.

Excited States in Single Crystalline Pentacene

The detailed examination of the PFP/PEN (001) single crystal interface was accompanied by a thorough analysis of the optical properties of neat PEN single crystals as reference. The investigation of the temperature dependent photoluminescence reveals four main emission bands as a result of different excitonic states in the crystal. These bands can be distinguished by their emission energy, as well as by their luminescence intensity as a function of temperature. Whereas three of the four detected emission bands show a phonon related non-radiative decay contribution, consistent with literature [He+05], the fourth emission band at 1.43 eV has not been reported so far. Its distinct temperature behavior shows a thermally activated increase in emission intensity with an activation energy of (9 ± 2) meV superimposed by a phonon associated non-radiative depopulation suggesting a trap state within the crystal or at its surface populated by thermally activated exciton diffusion.

Excited States in Zinc Phthalocyanine

Following the characterization of the donor-acceptor interfaces, the focus of the book shifts from the investigation of heterointerfaces to the photophysics inherently in-duced by the molecular packing in crystalline aggregates comprised of only one con-stituent, the molecule zinc phthalocyanine.

Single ZnPc molecules solved in DMSO were analyzed via UV/Vis absorption and photoluminescence spectroscopy as a reference for the complementary studies on the solid state. The electronic transitions were analyzed by means of a Franck-Condon model revealing two electronic transitions at 1.84 eV and 2.05 eV in the Q-band absorption region and are ascribed to a breaking of molecular symmetry according to Theisen *et al.* [The+15]. The photoluminescence occurs from the lowest excited state with the 0-0 transition centered at 1.83 eV. A change in the electronic-vibrational coupling between photon absorption and emission has been identified by a change in the Huang-Rhys parameters of the dominant vibrational modes.

The ZnPc α phase crystal structure [Erk04] is defined by a slip angle of 24.62° between two π-stacked molecules, which corresponds to an H-aggregate in a dipole-dipole interaction picture. The H-aggregate formation is corroborated by the red shifted absorption spectra and the large Stokes shift (390 meV) of the luminescence compared to the β phase J-aggregate. The magnitude of the Stokes shift together with the temperature dependent photoluminescence data are consistent with an excimer emission as the dominant radiative channel. The emission could be consistently described by a semi-classical model excimer [BKE68], as well as by its quantum mechanical extension explicitly developed in this book.

The ZnPc β phase unit cell [Luc+20] yields a π-stacking along the crystallographic [010] direction with a slip angle of 45.52° corresponding to a J-aggregate in a dipole-dipole interaction picture. The blue shift and the enhancement of the low energy absorption component, together with the small luminescence Stokes shift of 33 meV compared to the α phase, support this aggregate assignment. Above temperatures of 100 K, the emission is consistent with that of a Frenkel exciton polaron as the main emission channel. The observed vibrational modes are in agreement with the single molecule emission found in DMSO, but the vibronic coupling in the crystalline aggregate expressed by the Huang-Rhys parameter significantly changes in strength with temperature. Below 100 K, the observed temperature dependent luminescence is consistent with a superradiant emission caused by exciton delocalization along the ZnPc [010] stacking direction.

Phase Transition Kinetics and Dual Luminescent OLEDs

By virtue of the spectrally distinct luminescence, the kinetics of the thermally induced first order $\alpha \rightarrow \beta$ phase transition have been traced. The transition mechanism is identified as a polyhedral grain growth described by a Johnson-Mehl-Avrami-Kolmogorov model [Avr40]. Furthermore, a detailed examination of the growth kinetics reveal an Avrami exponent of 4.46 ± 0.02, indicating a non constant nucleation rate of the β phase during the phase transformation. Controlling the transition kinetics enabled the successful fabrication of dual luminescent OLEDs by tuning the volume ratio of both ZnPc polymorphs in the active layer. A full optoelectronic characterization confirms the long term stability of the diodes, a key criteria for technological application of such devices in modern thin film optoelectronics.

Low Temperature Electronic Phase Transition

ZnPc single crystals comprising the thermodyanimcally stable β phase crystal structure showed a sudden change in luminescence at 100 K. Combining polarization and temperature dependent spectrally resolved fluorescence measurements with an exciton polaron model [SY11], exciton delocalization is identified as the driving mechanism behind this electronic phase transition. Using a qualitative model based on the spatially anisotropic dipole-dipole interactions in the unit cell, two main coupling contributions have been identified to explain the temperature dependent change in emission polarization. The short molecular stacking distance of 4.853 Å along the crystallographic [010] direction favor a strong J-type exciton coupling along these molecular chains. The mirror symmetry of two neighboring chains leads to an overall unpolarized emission. At low temperatures the lack of phonon induced interference enables excitons on neighboring chains to couple, forming new emitting inter chain states. The radiative transition from the lowest excited state of these inter chain excitons is polarized perpendicular to the crystallographic [010] direction, thereby, explaining the strong polarization perpendicular to the macroscopic long crystal axis at low temperatures.

In summary, the results in this book show that by altering the molecular packing in crystalline aggregates and at interfaces the resulting photophysics of the system can be tuned reliably in a broad spectral range. This offers the opportunity of utilizing the deterministic tuning of the molecular packing in the active layer as an additional, highly reliable approach, besides the well established chemical modification of the constituents, to optimize the performance of future organic optoelectronic devices.

7. Zusammenfassung

Ziel dieser Arbeit war es, den Einfluss der zugrunde liegenden Kristallstruktur und der damit einhergehenden molekularen Anordnung auf die angeregten Zustände in molekularen Aggregaten zu untersuchen. Zu diesem Zweck wurden zwei Modellsysteme ausgewählt. Der optisch anregbare und detektierbare Ladungstransferzustand im Donor-Akzeptor Komplex Pentacen-Perfluoropentacen (PEN-PFP) und die Möglichkeit, hoch definierte kristalline Grenzflächen herzustellen, ermöglichten detaillierte Einblicke in die räumlich anisotrope Ausbildung des Ladungstransferzustands. Durch Ausnutzen der gewonnenen Erkenntnisse beim Design von Bauteilen auf Basis dieser Donor-Akzeptor Grenzflächen konnte gezeigt werden, wie wichtig die morphologische Kontrolle ist, um das Auftreten von Fallenzuständen in Zusammenhang mit solchen Ladungstransferprozessen zu minimieren und damit die elektronischen Bauteileigenschaften zu verbessern. Für Zinkphthalocyanin (ZnPc) und dem ihm eigenen Polymorphismus konnte der Einfluss der molekularen Packung auf angeregte Zustände untersucht werden, ohne die chemische Struktur zu verändern. Durch die Kombination von Strukturuntersuchungen, optischer Absorptions- und Emissionsspektroskopie und Franck-Condon Modellierungen wurde der Ursprung der Emission der angeregten Zustände in der strukturellen α und β Phase über einen großen Temperaturbereich von 4 K bis 300 K offen gelegt. Mithilfe der erlangten Einsichten wurde die Kinetik des $\alpha \rightarrow \beta$ Phasenübergangs erster Ordnung charakterisiert und zur Herstellung von dual-lumineszenten OLEDs verwendet.

Pentacen Kristallstruktur und Morphologie der PFP/PEN Grenzschicht

Als Grundlage für die spektroskopischen Untersuchungen zur Ausbildung von Ladungstransferzuständen, insbesondere für den Fall einer senkrecht aufeinander stehenden Orientierung der Moleküle, wurde eine grundlegende morphologische Charakterisierung der reinen PEN (001) Substratkristalle sowie der PFP/PEN Kristallgrenzfläche durchgeführt. Messungen der reinen PEN Einkristalle mittels Röntgenbeugung (XRD, engl.: *X-Ray diffraction*) bestätigen die für Sublimationskristalle erwartete Volumenkristallphase mit einer *out-of-plane* (engl.: aus der Ebene heraus, d.h. senkrecht zur Kristalloberfläche) (001) Gitterkonstanten von 14,1 Å [Mat+01]. Die Beugungsmuster weisen jedoch auch eindeutig auf die Bildung eines weiteren Polymorphs hin, da für mehrere Proben eine koexistierende zweite Strukturphase mit einem out-of-plane Gitterabstand von 14,3 Å gefunden wurde. Wahrscheinlich handelt es sich hierbei um die verbleibende Fraktion eines Polymorphs, das während des Kristallwachstums gebildet wurde. Die Rasterkraftmikroskopie (AFM, engl.: *atomic force microscopy*) der reinen Kristalloberfläche zeigt großflächige Terrassen, die durch monomolekulare Stufen getrennt sind und somit die sehr gute Probenqualität sowie die aufrechte Orientierung der PEN Moleküle an der Kristalloberfläche bestätigen. Zusätzliche phasensensitive AFM Messungen nach Abscheiden einer Submonolage PFP auf einen PEN (001) Kristall beweisen, dass bereits die PFP Moleküle direkt an der PFP/PEN (001) Kristallgrenzfläche ebenfalls aufrecht stehen. Zudem legen die XRD Untersuchungen von auf PEN Einkristallen gewachsenen PFP Dünnfilmen eindeutig ein PFP Wachstum in der strukturellen Dünnschichtphase nahe [Sal+08a].

Bildung von Ladungstransferzuständen an Donor - Akzeptor Grenzflächen

Die Bildung von Ladungstransferzuständen wurde anhand zweier grundlegender Konfigurationen der Molekülstapel untersucht: „flächig zugewandt" (*cofacial*) und „axial aufeinander stehend" (*head-to-tail*). Die *cofacial* Stapelung wurde mit PFP dotierten PEN Schichten und Heterostrukturdioden realisiert, die durch Koverdampfen der molekularen Konstituenten im PEN:PFP Verhältnis von 12:1 und 3:1 hergestellt wurden. Mittels UV/Vis-Absorptionsspektroskopie wurde ein in den reinen Materialien nicht vorhandenes Absorptionsband bei 1,57 eV gefunden, das laut Literatur die Bildung eines Ladungstransferzustandes bestätigt [Bro+11]. Die *head-to-tail* Anordnung des Donor-Akzeptor Komplexes wurde mittels PEN (001) Einkristallen als Substrat für das anschließende PFP Wachstum realisiert. Diese Grenzfläche wurde durch temperaturabhängige Photolumineszenzspektroskopie und durch differentielle Reflektionsspektroskopie (DRS) untersucht, womit zum ersten Mal gezeigt werden konnte, dass für senkrecht aufeinander stehende PEN-PFP Moleküle kein Ladungstransferzustand auftritt. Mit beiden Grenzflächenanordnungen wurden anschließend prototypische Bilagen-(*head-to-tail*) und Volumen-Heterostrukturdioden (*cofacial*) hergestellt. Die optoelektronische Charakterisierung der Dioden zeigte für beide Grenzflächengeometrien eine Elektrolumineszenzbande bei 1,4 eV, was der Emission des Ladungstransferzustands für flächig zugewandte Moleküle zugeordnet wird [Ang+12]. Jedoch führt bei ambipolarer Ladungsträgerinjektion die sehr viel größere Dichte an Ladungstransferzuständen in Volumen-Heterostrukturen zu einer geringeren Stromdichte in den entsprechenden Dioden, da die Ladungstransferkomplexe Fallenzustände für die Ladungsträgerpaare darstellen. Diese Ergebnisse zeigen, wie wichtig es ist, die molekulare Orientierung an Donor-Akzeptor Grenzflächen zu kontrollieren, um maximal effiziente Heterostrukturbauteile zu produzieren.

Excitonische Zustände in Pentacen Einkristallen

Die genaue Untersuchung der PFP/PEN (001) Einkristall Grenzfläche erforderte zunächst eine detaillierte Studie an reinen PEN Einkristallen als Referenz. Die Analyse der temperaturabhängigen Photolumineszenz der Volumenkristalle zeigte vier Hauptemissionsbeiträge, die auf verschiedene excitonische Zustände hinweisen. Diese sind aufgrund ihrer Emissionsenergie und ihrer temperaturabhängigen Lumineszenzintensität voneinander unterscheidbar. Während drei der vier Emissionsbänder in Übereinstimmung mit der Literatur einem durch Phononen bedingten nichtstrahlenden Zerfallskanal zugeordnet werden können [He+05], gibt es zur vierten Komponente bei 1,43 eV noch keine Berichte in der Literatur. Das ausgeprägte Verhalten mit der Temperatur zeigt einen temperaturaktivierten Anstieg der Emissionsintensität mit einer Aktivierungsenergie von (9 ± 2) meV, welchem eine von Phononen getriebene nichtstrahlende Entvölkerung überlagert ist. Dieses Verhalten legt einen emittierenden Fallenzustand, der im Volumen oder an der Kristalloberfläche lokalisiert ist und durch thermisch aktivierte Excitonendiffusion bevölkert wird, als Ursache für das beobachtete Emissionsband nahe.

Angeregte Zustände in Zinkphthalocyanin

Nach der Charakterisierung der Donor-Akzeptor Grenzflächen verlagerte sich der Schwerpunkt der Arbeit von der Untersuchung gemischter Grenzflächen auf die Untersuchung der durch die molekulare Packung bestimmten Photophysik in kristallinen Aggregaten, die aus nur einer einzigen chemischen Komponente, dem Zinkphthalocyaninmolekül, bestehen.

Einzelne ZnPc Moleküle, die in DMSO gelöst wurden, wurden mittels UV/Vis Absorptions- und Photolumineszenzspektroskopie analysiert, um als Referenz für die anschließenden Studien an Festkörpern zu dienen. Die elektronischen Übergänge wurden mit Hilfe eines Franck-Condon Modells simuliert. Die zwei detektierten elektronischen Übergänge im Q-Band der Absorption bei 1,84 eV und 2,05 eV wurden einem von Theisen *et al.* [The+15] beschriebenen Symmetriebruch im Molekül zugeordnet. Der elektronische 0-0 Übergang der Photolumineszenz aus dem niedrigsten energetischen Zustand tritt bei 1,83 eV auf. Zwischen Photonenabsorption und -emission zeigt sich eine Änderung in der elektronisch-vibronischen Kopplung anhand der jeweiligen Huang-Rhys Parameter der dominanten Schwingungsmoden.

In der α Phase liegt ZnPc in einer Kristallstruktur vor [Erk04], bei der die π-gestapelten Moleküle einen Versatzwinkel von 24,62° haben, was die Ausbildung eines H-Aggregats infolge der Dipol-Dipol-Wechselwirkung hervorruft. Dies wird durch eine dem J-Aggregat der β Phase gegenüber rotverschobene Absorption und durch die große Stokes Verschiebung von 390 meV nahegelegt. Die Größe der Stokes Verschiebung und die temperaturabhängige Photolumineszenz bestätigen die Excimer Emission als dominanten radiativen Zerfallskanal. Die Emission kann sowohl mit einem semiklassischen Excimer Modell [BKE68] als auch mit der in dieser Arbeit explizit entwickelten quantenmechanischen Erweiterung konsistent beschrieben werden.

Die ZnPc β Phase weist eine Einheitszelle auf [Luc+20], in der die Moleküle entlang der kristallographischen [010]-Richtung mit einem Versatzwinkel von 45,52° π-gestapelt sind. Dies führt aufgrund der Dipol-Dipol-Wechselwirkung zu einem J-Aggregat. Die Blauverschiebung und Erhöhung der niederenergetischen Absorptionskomponente im Vergleich zur α Phase, zusammen mit der kleinen Stokes Verschiebung der Lumineszenz von 33 meV, sprechen für diesen Aggregattyp. Für Temperaturen oberhalb von 100 K entspricht die Emission der eines Frenkel Exciton-Polarons. Die beobachteten Schwingungsmoden stimmen mit denen der Einzelmoleküle in Lösung überein, wobei sich die Kopplungen an den elektronischen Übergang, beschrieben durch die Huang-Rhys-Parameter, signifikant mit der Temperatur ändern. Unterhalb von 100 K kann die Temperaturabhängigkeit der Lumineszenz durch das Auftreten einer superradianter Emission durch Excitonendelokalisation entlang der ZnPc [010] Stapelrichtung konsistent erklärt werden.

Kinetik des Phasenübergangs und dual-lumineszente OLEDs

Aufgrund der spektral separierbaren Lumineszenzbeiträge kann die Kinetik des $\alpha \rightarrow \beta$ Phasenübergangs erster Ordnung analysiert werden. Der Übergangsmechanismus wird als polyedrisches Kornwachstum, das durch ein Johnson-Mehl-Avrami-Kolmogorov-Modell beschrieben werden kann, identifiziert [Avr40]. Des Weiteren ergab die genaue Analyse der Wachstumskinetik einen Avrami-Exponenten von 4,46 ± 0,02, was auf eine nichtkonstante Nukleationsrate der β Phase während des Phasenübergangs hindeutet. Die Kontrolle über die Übergangskinetik erlaubt die Herstellung von dual-lumineszenten OLEDs, in denen das Volumenverhältnis der beiden Polymorphe gezielt eingestellt wurde. Eine detaillierte optoelektronische Untersuchung der Dioden bestätigt deren Langzeitstabilität, eines der wichtigsten Kriterien für eine technologische Anwendung der Bauteile.

Elektronischer Phasenübergang bei tiefen Temperaturen

Die thermodynamisch stabile Kristallstruktur von ZnPc Einkristallen ist die β Phase. Die Kristalle zeigen bei 100 K eine plötzliche Änderung ihrer Lumineszenz. Mit polarisations- und temperaturabhängiger Fluoreszenzspektroskopie und einem Exciton-Polaron Modell [SY11] wurde hergeleitet, dass Excitonendelokalisation der verantwortliche Mechanismus für dieses optische Verhalten ist. Mit Hilfe eines qualitativen Modells, basierend auf der räumlich anisotropen Dipol-Dipol-Wechselwirkung in der Einheitszelle, wurden zwei Hauptmechanismen zur Erklärung der beobachteten Änderung in der Emissionspolarisation mit der Temperatur identifiziert. Der kleine Abstand von 4,853 Å zwischen den ZnPc Molekülen entlang der kristallographischen [010]-Richtung begünstigen eine starke J-artige Excitonenkopplung entlang dieser molekularen Ketten. Die Spiegelsymmetrie zweier benachbarter Ketten führt dann zu einer insgesamt unpolarisierten Emission. Bei niedrigen Temperaturen ermöglicht das Ausfrieren der Phononen eine Kopplung der excitonischen Zustände auf benachbarten Ketten. Die Emission aus dem energetisch niedrigsten Zustand der gekoppelten Kettenexcitonen ist senkrecht zur kristallographischen [010]-Richtung polarisiert und erklärt somit die beobachtete, starke Polarisation senkrecht zur makroskopisch langen Kristallachse.

Die Ergebnisse dieser book zeigen, dass durch die Änderung der molekula-ren Packung in kristallinen Aggregaten und an Grenzflächen die Photophysik der jeweiligen Anordnung gezielt variiert und über einen weiten Spektralbereich eingestellt werden kann. Durch gezieltes Einstellen der molekularen Orientierung in der aktiven Schicht eröffnet sich, neben den etablierten Möglichkeiten, wie zum Beispiel der chemischen Modifikation der molekularen Bausteine, eine zusätzlicher zuverlässiger Ansatz zur Optimierung der Leistung zukünftiger organischen, optoelektronischen Bauteilen.

Appendices

A. Dipole Approximation for Electronic Transitions

This section is complementary to the introduction to electronic transitions in section 2.3.1 and gives a deeper insight in the coming about of the transition rates and the Einstein coefficients. If not stated otherwise it is a summary of [HW03].

In [HW03] the following expression for a matrix element of an electronic transition from an initial state $|i\rangle$ to a final state $|f\rangle$ of a molecule caused by absorption or emittance of a photon with frequency ω_λ and polarization $\vec{e}_\lambda$ is deduced:

$$\mathcal{M}_{if} = \left\langle f \left| \sum_{j=1}^{N} \frac{-e}{m_0} \frac{1}{\sqrt{V}} \sqrt{\frac{\hbar}{2\epsilon_0\omega_\lambda}} \exp\left(i\vec{k}_\lambda\hat{r}_j\right) \vec{e}_\lambda\hat{p}_j \right| i \right\rangle \tag{A.1}$$

Here, e is the elementary charge, m_0 the electron's mass and V is an auxiliary volume used to quantize the light field. The matrix element (A.1) is the molecular part of the perturbation. The matrix element of light field's transition is not elaborated here. Concentrating on energies for electronic transitions, only the interaction of the molecule's N electrons with the light field are considered. The energies of the nucleis' vibrational states, which are included in the full perturbation Hamiltonian from (2.29) are not in resonance with the incident (or emitted) light, giving the matrix element a value of zero.

For the dipole approximation, two assumptions are made. First, it is assumed that the wavelength of the incident light λ is large compared with the molecule's size. This means that the expansion of the exponential term in (A.1) leads to [Fli18]

$$\exp\left(i\vec{k}\hat{r}\right) = 1 + i \underbrace{\vec{k}\hat{r}}_{"r/\lambda\approx0"} + ... \approx 1 \tag{A.2}$$

if the molecule's center of mass is chosen as the coordinate system's origin. Second, it is assumed that only one electron is causing the evolution from the initial to the final state. This implies that the sum in equation (A.1) can be dropped leading to the following expression for the matrix element:

$$\mathcal{M}_{if} = \frac{-e}{m_0}\sqrt{\frac{\hbar}{2\epsilon_0\omega_\lambda V}} \cdot \langle f|\vec{e}_\lambda \hat{p}_j|i\rangle \tag{A.3}$$

In many cases, it is more useful to evaluate equation (A.3) in space representation rather than in momentum representation, as the wave functions of molecular orbitals are given as functions of real space coordinates, for example to make use of the molecule's symmetries in the calculation. This introduces the well know *transition dipole moment* $\vec{M}_{if} = \langle f|\hat{\mu}|i\rangle$ with the dipole operator $e\hat{r}$.

The following relation for the commutator of the position operator $\hat{x}$ and the Hamiltonian $\hat{H}_0 = \hbar^2/2m\partial^2/\partial x^2$ is useful. The following calculations are performed in one dimension for clarity, but can easily be expanded to any number of dimensions. Let $f(x)$ be a suitable wave function, *i.e.* non zero, differentiable *etc.* Applying the commutator $[\hat{H}_0, \hat{x}]$ leads to the following relation

$$[\hat{H}_0, \hat{x}]f(x) =$$
$$= \hat{H}_0\hat{x}f(x) - \hat{x}\hat{H}_0 f(x)$$
$$= -\frac{\hbar^2}{2m}\left(\frac{\partial^2}{\partial x^2}\left(xf(x) - xf''(x)\right)\right)$$
$$= -\frac{\hbar^2}{2m}\left(f'(x) + f'(x) + xf''(x) - xf''(x)\right)$$
$$= -\frac{\hbar^2}{m}\frac{\partial}{\partial x}f(x) = \frac{1}{m}\frac{\hbar}{i}\underbrace{\frac{\hbar}{i}\frac{\partial}{\partial x}}_{\hat{p}}f(x)$$

$$[\hat{H}_0, \hat{x}]f(x) = \frac{1}{m}\frac{\hbar}{i}\hat{p}f(x), \tag{A.4}$$

yielding

$$[\hat{H}_0, \hat{x}] = \frac{1}{m}\frac{\hbar}{i}\hat{p}. \tag{A.5}$$

Inserting (A.5) in (A.3) together with $\hat{H}_0|i\rangle = E_i|i\rangle$ and $\hat{H}_0|f\rangle = E_f|f\rangle$ yields

$$\mathcal{M}_{if} = \frac{-e}{m_0}\sqrt{\frac{\hbar}{2\epsilon_0\omega_\lambda V}}\frac{im_0}{\hbar}\vec{e}_\lambda\langle f|[\hat{H}_0, \hat{r}]|i\rangle$$
$$= -\frac{i}{\hbar}\sqrt{\frac{\hbar}{2\epsilon_0\omega_\lambda V}}\vec{e}_\lambda e\langle f|\hat{H}_0\hat{r} - \hat{r}\hat{H}_0|i\rangle$$
$$= -\frac{i}{\hbar}\sqrt{\frac{\hbar}{2\epsilon_0\omega_\lambda V}}\vec{e}_\lambda e\left(\langle f|\hat{H}_0\hat{r}|i\rangle - \langle f|\hat{r}\hat{H}_0|i\rangle\right)$$

$$= -\frac{i}{\hbar}\sqrt{\frac{\hbar}{2\epsilon_0\omega_\lambda V}}\,\vec{e}_\lambda e\big(\underbrace{\langle H_0 f|\,\hat{r}\,|i\rangle}_{\substack{\hat{H}_0 \text{ hermite}\\ \langle H_0 f|=E_f\langle f|}} - \langle f|\,\hat{r}\,\underbrace{|H_0 i\rangle}_{E_i|i\rangle}\big)$$

$$\mathcal{M}_{if} = -i\sqrt{\frac{\hbar}{2\epsilon_0\omega_\lambda V}}\,\vec{e}_\lambda\,\frac{E_f - E_i}{\hbar}\,\langle f|\,e\hat{r}\,|i\rangle. \tag{A.6}$$

Let the respective photon be in resonance with the energy gap, than

$$\omega_\lambda = \omega_{if} = \frac{E_f - E_i}{\hbar}. \tag{A.7}$$

Together with the definition of the transition dipole moment $\vec{M}_{if} = \langle f|e\hat{r}|i\rangle$ equation (A.6) can be expressed as

$$\mathcal{M}_{if} = -i\sqrt{\frac{\hbar\omega_{if}}{2\epsilon_0 V}}\,\vec{e}_\lambda\vec{M}_{if}. \tag{A.8}$$

To determine the transition rates using Fermi's golden rule (2.31) the matrix element of the light field as well as the potential degeneracy of the final states has to be considered. In [HW03] this yields the following relations:

$$\text{spontaneous emission:}\quad W_{21} = \frac{2\pi}{\hbar}\,\underbrace{\frac{V}{(2\pi)^3}\frac{1}{(\hbar c)^3}\,(\hbar\omega_{if})^2}_{\text{DOS } D(\omega_{if})}\,|\mathcal{M}_{if}|^2\,d\Omega \tag{A.9a}$$

$$\text{absorption:}\quad W_{12} = \frac{2\pi}{\hbar}\,\underbrace{\frac{n\,\hbar\omega_{if}}{\Delta\omega\,d\Omega\,V}\frac{V}{\hbar^2\omega_{if}}}_{\text{spectral energy density } \rho(\omega_{if},d\Omega)}\,|\mathcal{M}_{if}|^2\,d\Omega \tag{A.9b}$$

In (A.9a) c is the speed of light and the photon's density of states (DOS) $D(\omega)$ evaluated at the transition frequency ω_{if} in the solid angle $d\Omega$ considers all possible final states. The photon absorption (A.9b) is dependent on the photon density in the initial state. This is represented by the spectral energy density $\rho(\omega_{if}, d\Omega)$ at the respective transition frequency ω_{if}, given by the number of photons n with the energy $\hbar\omega_{if}$ normalized to the unit frequency interval $\Delta\omega$, the solid angle $d\Omega$ and the respective (auxiliary) volume V. Using the dipole approximation (A.8) for the matrix element $\mathcal{M}_{if}$ the transition rates (A.9a) and (A.9b) can be expressed as

$$\text{spontaneous emission:}\quad W_{21} = \underbrace{\frac{1}{8\pi^2 c^3 c_0}\,\omega_{if}^3\,|\vec{e}_\lambda\vec{M}_{if}|^2\,d\Omega}_{A_{21}}, \tag{A.10a}$$

$$\text{absorption:}\quad W_{12} = \rho(\omega_{if}, d\Omega)\,\underbrace{\frac{\pi}{\hbar^2\epsilon_0}\,|\vec{e}_\lambda\vec{M}_{if}|^2\,d\Omega}_{B_{12}}. \tag{A.10b}$$

B. Excimers

B.1. Displaced Harmonic Oscillator Simulation

In the excimer description in section 2.3.2 simulated excimer emission spectra are shown in figure 2.13 b) on page 42. The simulation procedure as well as the used parameters will be stated, followed by a short discussion on the validity of the model.

The emission spectrum for a displaced harmonic oscillator simplifies from equation (2.49) to

$$\mathcal{W}(E;T) = \sum_j \sum_k P(k \cdot E_{\text{vib}}, T) \left| \langle j \mid k \rangle \right|^2 \delta \left(E - E_{kj} \right) \tag{B.1}$$

where E_{vib} is the energy distance of the vibronic states in both oscillators and E_{kj} is the energy distance between the k-th vibronic level in the excited state and the j-th vibronic level of the ground state. In the displaced oscillator model some transitions are degenerated. For example all transitions for which the vibronic quantum number does not change ($n \to n$) have the same energy as the $0 \to 0$ transition. In general the transitions $n \to m$ and $n' \to m'$ are degenerated if

$$|n - m| = |n' - m'| =: \Delta v \tag{B.2}$$

with the corresponding transition energy $E_{\Delta v}$. The intensity at the energy $E_{\Delta v}$ is given as

$$\delta \left(E - E_{\Delta v} \right) \tilde{\mathcal{W}} \left(E_{\Delta v}; T \right) = \delta \left(E - E_{\Delta v} \right) \sum_{|m-n|=\Delta v} P(m \cdot E_{\text{vib}}, T) \left| \langle n \mid m \rangle \right|^2 . \tag{B.3}$$

The respective Franck-Condon factor can be calculated with (2.39). The probability of a vibronic state $|m\rangle$ with energy $E_m = (m + 1/2) E_{\text{vib}}$ being occupied at a temperature T is given by the Boltzmann distribution

$$P(E_m; T) = \frac{1}{Z} \exp\left(-\frac{m \cdot E_{\text{vib}}}{k_b T} \right) \tag{B.4}$$

with the canonical partition function

$$Z = \sum_z \exp\left(-\frac{E_z}{k_b T} \right) \tag{B.5}$$

and the Boltzmann constant k_b [Fli10]. For a numerical simulation of the spectra a suitable number of vibronic states for the excimer (N) and ground state (M), the Huang-Rhys parameter S as well as the vibronic energy quanta E_{vib} have to be chosen. With these parameters the transition intensities for each emission energy $E_{\Delta v}$ can be calculated from (B.3) and the full spectra is than given by

$$W(E; T) = \sum_{\Delta v} \delta (E - E_{\Delta v}) \tilde{W} (E_{\Delta v} ; T). \qquad (B.6)$$

Assuming a gaussian emission with an energetic disorder $\sigma_E = 10\,\text{meV}$ and convoluting the right side of equation (B.6) the simulated emission spectra is obtained. The chosen parameters for the simulation of figure 2.13 b) are listed in table B.1.

Parameter	Value
Vibronic energy quanta E_{vib}	12 meV
Huang-Rhys parameter S	8
Vibronic levels excited state N	9 (1+8)
Vibronic levels ground state M	21 (1+20)
Temperature	4 K, 150 K, 300 K

Table B.1.: Parameters for simulating the excimer emission based on a displaced harmonic oscillator model for figure 2.13 b).

The vibrational energy quanta is chosen to reflect the usual phonon energy in organic solids. For two displaced, otherwise equal harmonic oscillators, the Huang-Rhys parameter gives the "mean" quantum number of the energy a classical harmonic ground state oscillator would have at equal displacement. Usually the final states of the excimer fluorescence have high quantum numbers [Bir75] and the Huang-Rhys parameter S was chosen to reflect that. The $N = 9$ vibronic levels of the excimer (1 ground state, 8 excited states) were chosen so that all relevant thermal occupation at $T = 300\,\text{K}$ was captured. The arbitrary cutoff m_{co} was at $P(E_{m_{\text{co}}}, 300\,\text{K}) < 0.01$. The real ground state potential is not harmonic for high quantum numbers which leads to a reduction of energetic distance between two subsequent states before reaching the dissociation energy. The lowest emission energy has a sharp cutoff as the emission is caused by transitions from the vibronic ground state into the quasi continuum just below the dissociation energy. This leads to a high energy flank for higher temperatures in excimer emission. The sharp low energy cutoff was accounted for by limiting the ground state vibrational states to $M = 20$ which leads to a slight asymmetry at the high energy flank for 300 K.

B.2. Semiclassical Harmonic Oscillator Approximations

In this section, the analytical expression of the line shape functions of the individual emission transition spectra of the excited state oscillator levels $P_n(E)\mathrm{d}E$ necessary to express equation (5.13) are derived for the ground and the first five excited states.

The starting point is equation (5.12) relating the excited state oscillator wave function to the emission profile:

$$P_n(E)\mathrm{d}E = \left|\Psi_n(q(E))\right|^2 \frac{\mathrm{d}q}{\mathrm{d}E}\mathrm{d}E \tag{B.7}$$

The wave function can be described by a harmonic oscillator wave function ψ_n at q_e as $\Psi_n(q) = \psi_n(q - q_e)$. The energy of the initial state ψ_n is given by

$$D_{\mathrm{E}n} = D_\mathrm{e} + (2n+1)E_0 \tag{B.8}$$

yielding a photon energy as a function of inter molecular coordinate of

$$E_n(q) = D_{\mathrm{E}n} - R(q) = D_{\mathrm{E}n} - R_0 q^2. \tag{B.9}$$

The inverse function gives the inter molecular coordinate as a function of photon energy E as

$$q_n(E) = \sqrt{\frac{D_{\mathrm{E}n} - E}{R_0}} \tag{B.10}$$

and the associated Jacobian is given by

$$\frac{\mathrm{d}q_n}{\mathrm{d}E} = \frac{1}{2\sqrt{R_0\left(D_{\mathrm{E}n} - E\right)}}. \tag{B.11}$$

Defining the oscillator constant

$$\alpha = \frac{\mu\omega}{\hbar} = \frac{2\mu E_0}{\hbar^2} \tag{B.12}$$

and a generalized oscillator coordinate for each level n

$$\tilde{q}_n(E) = \sqrt{\frac{D_{\mathrm{E}n} - E}{R_0}} - q_e \tag{B.13}$$

as a function of energy the wave function of the oscillator is expressed as

$$\psi_n(\tilde{q}_n) = \left(\frac{\alpha}{\pi}\right)^{\frac{1}{4}} \frac{1}{\sqrt{2^n n!}} H_n(\sqrt{\alpha}\,\tilde{q}_n) e^{-\frac{1}{2}\alpha \tilde{q}_n^2} \tag{B.14}$$

with the respective hermite polynom $H_n(x)$. Evaluating the wave function (B.14) using (B.12) and (B.13) and substituting the results in (B.7) yields the following expressions for $n \in \{0,\dots,5\}$:

$$P_0(E)\mathrm{d}E = \frac{1}{2}\sqrt{\frac{\alpha}{R_0\pi\left(D_{E_0}-E\right)}}\,\exp\left(-\alpha\tilde{q}_0^2\right)\mathrm{d}E \tag{B.15}$$

$$P_1(E)\mathrm{d}E = \sqrt{\frac{\alpha}{R_0\pi\left(D_{E_1}-E\right)}}\,\alpha\tilde{q}_1^2\,\exp\left(-\alpha\tilde{q}_1^2\right)\mathrm{d}E \tag{B.16}$$

$$P_2(E)\mathrm{d}E = \frac{1}{4}\sqrt{\frac{\alpha}{R_0\pi\left(D_{E_2}-E\right)}}\,\left(2\alpha\tilde{q}_2^2-1\right)^2\,\exp\left(-\alpha\tilde{q}_2^2\right)\mathrm{d}E \tag{B.17}$$

$$P_3(E)\mathrm{d}E = \frac{1}{6}\sqrt{\frac{\alpha}{R_0\pi\left(D_{E_3}-E\right)}}\,\left(2\alpha\tilde{q}_3^2-3\right)^2\alpha\tilde{q}_3^2\,\exp\left(-\alpha\tilde{q}_3^2\right)\mathrm{d}E \tag{B.18}$$

$$P_4(E)\mathrm{d}E = \frac{1}{48}\sqrt{\frac{\alpha}{R_0\pi\left(D_{E_4}-E\right)}}\,\left(4\alpha^2\tilde{q}_4^4-12\alpha\tilde{q}_4^2+3\right)^2\,\exp\left(-\alpha\tilde{q}_4^2\right)\mathrm{d}E \tag{B.19}$$

$$P_5(E)\mathrm{d}E = \frac{1}{120}\sqrt{\frac{\alpha}{R_0\pi\left(D_{E_5}-E\right)}}\,\left(\left(2\alpha\tilde{q}_5^2-5\right)^2-10\right)^2\alpha\tilde{q}_5^2\,\exp\left(-\alpha\tilde{q}_5^2\right)\mathrm{d}E \tag{B.20}$$

B.3. Global Fitting Procedure for Franck-Condon Excimer Model

The residual function is built concatenating the difference between the emission data and respective transition spectrum (equation (5.16)) multiplied with an amplitude $I_0(T)$ for each temperature. This yields a residual set function

$$R\left(D_e, \sigma, E_{0;\mathrm{EX}}; I_0(T_1), E_{0;\mathrm{GS}}(T_1), q_e(T_1), \dots, I_0(T_n), E_{0;\mathrm{GS}}(T_n), q_e(T_n)\right) =$$
$$\left\{r_{T_1}\left(D_e, \sigma, E_{0;\mathrm{EX}}; I_0(T_1), E_{0;\mathrm{GS}}(T_1), q_e(T_1)\right); \dots; r_{T_n}\left(D_e, \sigma, E_{0;\mathrm{EX}}; I_0(T_n), E_{0;\mathrm{GS}}(T_n), q_e(T_n)\right)\right\} \tag{B.21}$$

with

$$r_{T_j}\left((D_e, \sigma, E_{0;\mathrm{EX}}; I_0(T_j), E_{0;\mathrm{GS}}(T_j), q_e(T_j)\right) =$$
$$\bar{I}_{\mathrm{obs}}(T_j; E_1) - I_0(T_j)\cdot\bar{I}\left(E_1; T_j, q_e(T_j), E_{0;\mathrm{EX}}, E_{0;\mathrm{GS}}(T_j), D_e, \sigma\right); \dots;$$
$$\bar{I}_{\mathrm{obs}}(T_j; E_m) - I_0(T_j)\cdot\bar{I}\left(E_m; T_j, q_e(T_j), E_{0;\mathrm{EX}}, E_{0;\mathrm{GS}}(T_j), D_e, \sigma\right) \tag{B.22}$$

with $\bar{I}_{\mathrm{obs}}(T_j; E_i)$ as the observed transition spectrum at the temperature T_j and emission energy E_i. The set R comprises entries resembling the difference between observed emission at energy E_i for each temperature T_j and intensity distribution of the overall intensity $I_0(T_j)$ by the Franck-Condon factors summerized in $\bar{I}$. Minimizing this function by a least square approach within the *LMFIT* python package [New+14] by variation of the given parameter set $\{D_e, \sigma, E_{0;EX}, I_0(T_1), q_e(T_1), E_{0;GS}(T_j), \ldots, I_0(T_n), q_e(T_n), E_{0;GS}(T_n)\}$ leads to the presented results in section 5.4.3. The parameters D_e, σ, and $E_{0;EX}$ are global parameters, *i.e.* are set as a constant value for all temperatures, while $I_0(T_j)$, $q_e(T_j)$, and $E_{0;GS}(T_j)$ are individual parameters for each temperatures.

C. Perfluoropentacene growth on Pentacene crystals

The following profile line analysis of a 20 nm PFP layer grown on a PEN crystal has been published alongside [Ham+20] as supplementary information as figure S2. The alongside published text and the figure are shown below without change, except the adjustment of the figure labeling and the change of citation style:

"Figure [C.1] depicts an atomic force microscopy (AFM) image of the 20 nm PFP film on top of the PEN (001) crystal surface comprising oriented, needle-like crystallites. In figure [C.1] (b) a high-resolution blowup of the region of interested marked by the white square in (a) is depicted. Three height profiles, marked by horizontal lines in the AFM image, have been analyzed by using the terrace fitting tool of the Gwyddion software package [[NK12]] under the assumption of a fixed step height and using a polynom of first order to compensate for the tilting of the individual grains. The profiles together with the respective polynomial fits are shown in figure [C.1] (c), (d) and (e) and are consecutively numbered according to the AFM image. The resulting step heights, as indicated in the related graphs, agree well to the (100) lattice constant determined determined by X-ray diffraction as described above as well as in the main text."

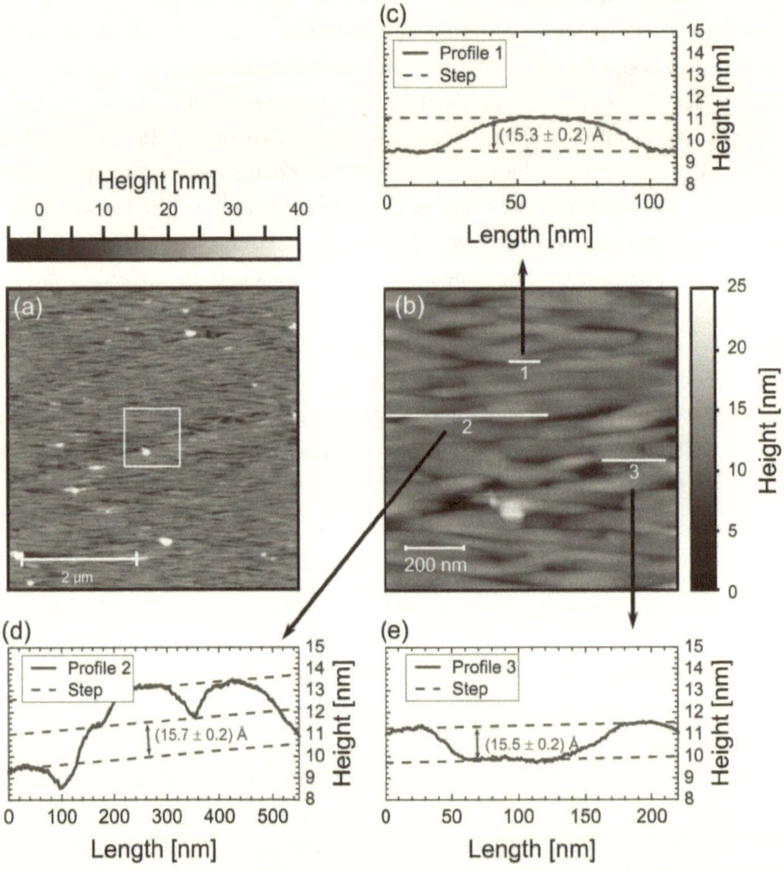

Figure C.1.: (a) AFM image of a 20 nm thick PFP thin film grown on top of a PEN (001) single-crystal surface facet. b) High resolution AFM image of marked area in (a) (white square) with selected line profiles 1, 2, and 3 indicated by white lines. The profiles 1, 2, and 3, are plotted in (c), (d) and (e) respectively including the estimated monomolecular step heights. Previously published as supplementary information alongside [Ham+20]

D. Zinc Phthalocyanine

D.1. Three Mode Absorption Fit

The absorption spectra of ZnPc can be modeled under the assumption of three vibronic modes. The transition spectrum in a displaced harmonic oscillator model is given according to (2.38) as

$$\overline{I}(E) = \sum_{m} \sum_{i_m=0}^{5} \Gamma\left(E; E_0 + i_\alpha \cdot E_{\mathrm{vib},\alpha} + i_\beta \cdot E_{\mathrm{vib},\beta} + i_\gamma \cdot E_{\mathrm{vib},\gamma}, \sigma\right) \prod_{k=\alpha,\beta,\gamma} \frac{e^{-S_k}}{i_k!} S_k^{i_k} \qquad (D.1)$$

with $m \in \{\alpha, \beta, \gamma\}$ indexing the three different vibronic modes and Γ as a Gauss function given by equation (5.2). Figure D.1 shows the resulting fit curve and the stick spectra labeled according to the respective vibrational final state. Resulting fitting parameters are listed in table 5.1.

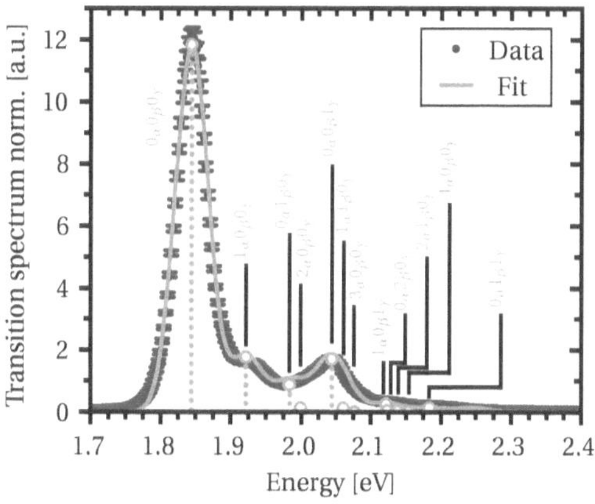

Figure D.1.: Absorption and three mode Franck-Condon fit of ZnPc in DMSO

E_0			σ		
1.844 eV			(23.3 ± 0.1) meV		
$E_{\mathrm{vib},\alpha}$	S_α	$E_{\mathrm{vib},\beta}$	S_β	$E_{\mathrm{vib},\gamma}$	S_γ
(78 ± 1) meV	0.149 ± 0.003	(139 ± 2) meV	0.075 ± 0.003	(200 ± 1) meV	0.145 ± 0.003

Table D.1.: Fit parameters obtained from three mode Franck-Condon fit given by equation (D.1).

D.2. DMSO Raman Spectrum

The Raman signal of a neat DMSO sample measured as a solvent spectra (*c.f.* section 3.4.2) with 685 nm excitation is shown in figure D.2. Table D.2 lists the observed relative wavenumber $\Delta\overline{\nu}$ (reading accuracy) for the respective peak label (ID) and the corresponding vibrational mode assigned from [HC61]

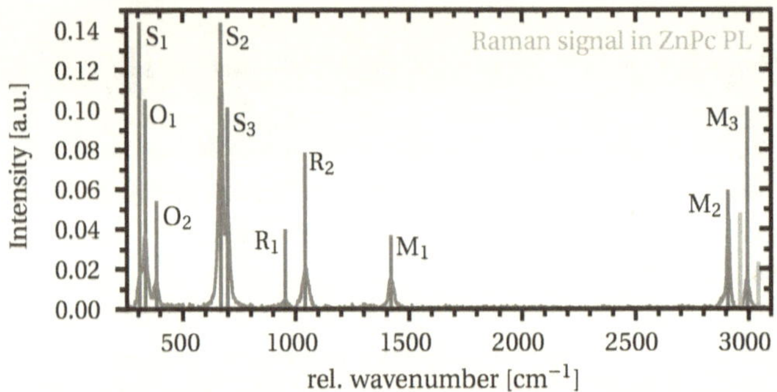

Figure D.2.: Raman spectrum of DMSO with 685 nm excitation. The cyan lines mark position of Raman signal observed in 1×10^{-8} mll^{-1} and 1×10^{-7} mll^{-1} ZnPc-DMSO samples.

ID	$\Delta\overline{\nu}$[cm^{-1}]	vibrational mode [HC61]
S$_1$	311 ± 3	C–S–C deformation
S$_2$	666 ± 3	symmetric C–S stretch
S$_3$	696 ± 3	asymmetric C–S stretch
O$_1$	335 ± 3	asymmetric C–S–O deformation
O$_2$	382 ± 3	symmetric C–S–O deformation
R$_1$	953 ± 5	Methyl rock
R$_2$	1044 ± 3	
M$_1$	1418 ± 5	degenerate C–H deformation
M$_2$	2909 ± 3	symmetric C–H stretch
M$_3$	2994 ± 4	degenerate C–H stretch

Table D.2.: Fit parameters obtained from three mode Franck-Condon fit given by equation (D.1).

D.3. Polarization Dependency of Low Temperature 0-0 Transition

Figure D.3 shows the 0° and 80° polarization emission spectra of a ZnPc crystal at 5 K. In a) the full spectra are shown, revealing a large decrease in overall intensity. The two components of the 0-0 transition show a slightly different behavior with polarization. While the higher energy component seems to almost completely vanish, a residual peak is observed for the low energy peak. This is highlighted by figure D.3 b) in which the 0-0 transitions for both polarizations are shown normalized to their maximum intensity.

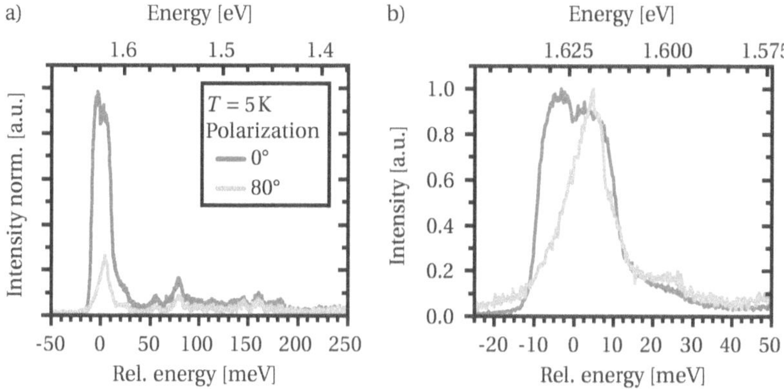

Figure D.3.: 0° and 80° polarized ZnPc crystal emission at 5 K. In a) both spectra are fully shown while in b) a blowup of the 0-0 transition of both polarizations normalized to the emission maximum is shown.

D.4. Estimation of Angles Between Two ZnPc Molecules

To estimate the angle between to adjacent molecular transition dipole moments located on ZnPc molecules within the crystallographic ($\bar{1}01$) of the ZnPc β polymorph the angle measurement tool of the software package Mercury 4.0 [Mac+20] is used. For this Purpose the direction of the molecular transition dipole moment is approximated by the conjuncture of the central Zn atom and the the C atom of the inner C–N ring closest to the intersection with the crystallographic ($\bar{1}01$) plane. The geometry is shown in figure D.4. The angle enclosed by the C_1 atom, the central Zn atom, and the central Zn atom of an adjacent ZnPc molecule is measured by the Mercury 4.0 angle measurement tool and is used for the dipole-dipole calculations.

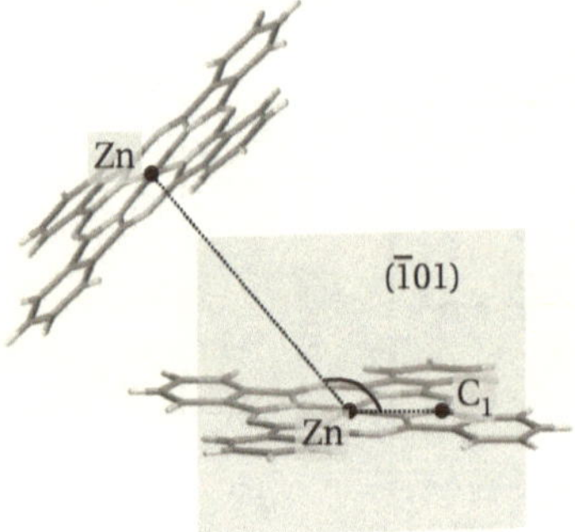

Figure D.4.: Estimation of angles between two ZnPc molecules in the ($\bar{1}01$) crystal plane. The three points used to determine the angles between two adjacent molecular transition dipole moments are the positions of the two center Zn atoms and the C atom (here labeled by C_1 of the inner C–N ring closest to the intersection with the crystallographic ($\bar{1}01$) plane (red).

www.ingramcontent.com/pod-product-compliance
Lightning Source LLC
LaVergne TN
LVHW040009200726
843493LV00005B/1193